THE COLOR ATLAS OF
GALAXIES

James D. Wray

The right of the
University of Cambridge
to print and sell
all manner of books
was granted by
Henry VIII in 1534.
The University has printed
and published continuously
since 1584.

Cambridge University Press

Cambridge

New York   New Rochelle   Melbourne   Sydney

Published by the Press Syndicate of the University of Cambridge
The Pitt Building, Trumpington Street, Cambridge CB2 1RP
32 East 57th Street, New York, NY 10022, USA
10 Stamford Road, Oakleigh, Melbourne 3166, Australia

First published 1988

Printed in Great Britain by W. S. Cowell Ipswich

British Library cataloguing in publication data

Wray, James D.
The color atlas of galaxies.
1. Galaxies – Atlases
I. Title
523.1′12   QB857

Library of Congress cataloguing in publication data

Wray, James D.
The color atlas of galaxies.
1. Galaxies – Atlases. I. Title.
QB857.W73   1988   523.1′12′0222   86–13715

ISBN 0 521 32236 7

Contents

Acknowledgments

I wish to express my deep appreciation to Harlan Smith, Director of the McDonald Observatory, for his encouragement and support for this work from its inception, without which the large amount of telescope time required, extending over nearly a decade before coming to fruition, could never have been realized. In addition he saw to it that I was provided with the necessary materials and much of the laboratory space required to convert the observations to color images. At least as responsible for the completion of this work was my wife, Elizabeth. Not only did she assist me at the telescope on many occasions, but she cheerfully encouraged my devotion to this project, from beginning to end. Without the support of these two individuals this work simply could not have been accomplished.

I would also like to express my appreciation for the support and assistance provided by the administrators and staffs of the three observatories which made their facilities available to me for this work. At the McDonald Observatory Tom Barnes, Chuck Cobb, David Doss, Ed Dutchover and Joe Liddell each put in many hours in support of this work, as did many others there, both directly and indirectly, including Curt Laughlin, superintendent, Jim Blocker, George Grubb, John Jordan, Pat Olivias and Martin Villareal. At the Cerro Tololo Inter-American Observatory in Chile, the support of Victor Blanco and John Graham was most appreciated. I am indebted to Halton Arp, George Preston and members of the Las Campanas Observatory scheduling committee for very generous awards of observing time at that observatory in Chile, without which the southern hemisphere galaxies would have been seriously underrepresented in the atlas. I wish to thank Ljubomir Papic, mountain superintendent at Las Campanas Observatory, together with Oscar Duhalde, Bill Robinson, Hernan Solis, Fernando Peralta and Angel Guerra and all the rest of the mountain staff for their generous assistance and hospitality during my visits there. I also wish to acknowledge the special efforts of Peter Wizinowich of the David Dunlap Observatory Southern Station on Las Campanas who devoted his holiday time to assisting me throughout my entire CTIO observing run as well as during much of my observing on Las Campanas. His efforts doubled the productivity of the CTIO run and considerably increased the number of observations I was able to accomplish at Las Campanas.

Many individuals of the department of astronomy at the University of Texas have either participated, assisted, or supported me in some way in this work. Among these are Frank Bash, G. Fritz Benedict, Karl Henize (Astronaut Office, Johnson Space Center), and Gregory Shields. I was assisted in the darkroom by Ray Lazenby and Celso Rodriguez, both of whom made major contributions to this work. I wish to thank a number of then graduate students, especially Ron Buta and Michael Newberry for their major assistance, particularly at the telescopes, and also Burt Beardsley and Roy Wellington. I wish to give very special thanks to Harold G. Corwin, to whom I am particularly indebted, not only for his assistance with the data from the Second Reference Catalog of Bright Galaxies, his verification of my identifications and determination of north for every field in the atlas, and his proof-reading of the manuscript, but also for his photoelectric photometry of the brightest stars in the atlas specifically for the purpose of enhancing the interpretive value of the color images throughout the atlas.

I also wish to thank Frank McLaughlin of Eastman Kodak for many helpful discussions regarding the Kodak Dye-Transfer Process, and Harry Heckathorn of the U.S. Naval Observatory for the loan of very useful equipment for this program.

I am indeed fortunate to have this opportunity to express my gratitude to two individuals whom I have admired and respected for many years, Allan Sandage and Gerard de Vaucouleurs. I am much indebted to both of them for their valuable comments and helpful admonitions regarding the atlas text. The influence each of these two giants in extragalactic research has had on my own thinking has, I hope, resulted in the creation of a kind of common space or perspective in this work. Not one which repeats, or even very much reflects, their views on problems of extragalactic research, mostly because I have been less than adequate as a student, but rather one in which they may feel a sense of unity midst much diversity.

The Foreword which Allan Sandage has so very kindly written for this atlas presents an element of historical perspective I could never have offered, and at the same time sets a tone for the atlas which is certain to stimulate the interest of many, and to arouse the curiosity of those who yet doubt the scientific utility of color images of galaxies.

Finally, I wish to thank Mary Wray who truly initiated this work when she gave me my first lesson in astronomical photography nearly forty years ago.

This material is based on work supported in part by the National Science Foundation under Grant No. AST 80–02112. Any opinions, findings, conclusions or recommendations expressed in this publication are those of the author and do not necessarily reflect the views of the foundation.

J.D.W.

Foreword

In this Color Atlas of Galaxies, Dr Wray has produced a work of fundamental importance for science as well as one of striking beauty. Here is a work which can be fully appreciated by many of its readers entirely through the sense of wonder and awe it inspires by its beautiful portrayal of these galaxies, yet one which offers at the same time the rational basis necessary for scientific application.

The scientific value of the work lies in two properties of the photographs which Wray meticulously maintained throughout his application of the Eastman three-color dye transfer process. First, the colors are true. A correct color balance was maintained consistently at all brightness levels. Second, the correct relation of surface brightness of any one galaxy to another in the Atlas was also achieved. These two properties permit the book to be, at the same time, a primary source of information for the professional scientist in choosing new research projects to advance the subject of observational cosmology, and an important pedagogical tool in the classroom to teach three central themes concerning galaxies (the Hubble classification system, so fundamental to the subject, Baade's population concept, and the facts of the life cycles of the individual stars that make up the stellar content of galaxies).

The second point needs explaining because it contains within it nearly the whole of galaxy research since 1916. These color photographs make plain what was found by a series of hypotheses, tested by difficult and often brilliant observations, and a grand synthesis by astronomers working from 1910 to around 1950 with black and white photographs (to be sure with different color sensitivities so as to be able to separate any given galaxy into different components). This early work was necessarily done in a most painstaking way before composite color images were available such as are in this book. But further, the color information could not then be interpreted in terms of age of dominant stars until methods of age-dating the stars and an understanding of the Hertzsprung-Russell diagram, explained by Wray in the introductory text, was achieved in the mid 1950's.

In 1916, Fredric Seares at the Mount Wilson Observatory discovered that the central parts of 'nebulae' (not then known as separate stellar systems) were much redder than the outlying areas that contain the spiral arms. This color difference between the inner and the outer regions of spiral galaxies is now known to be quite fundamental for reasons explained by Wray, and shown so well by the photographs of Sb, Sc, Sd and the Magellanic Cloud type galaxies in this Atlas. Not only is there a color difference, but the relative size of the red nuclear bulge and the blue spiral features changes as one proceeds along the Hubble sequence. Hubble's major paper which gave this central theme of his classification system was published in 1926. In it he emphasized that the larger the central nuclear bulge, the less pronounced are the spiral arms. In the limit of the sequence where the galaxies have no arms, one has either the diskless E galaxies or the disk S0 galaxies with no recent star formation, hence the color is red everywhere. At the opposite end of the sequence in the Sc to Sm types, the central bulge is reduced to a small or non-existent nucleus, while the spiral arms cover the face of the galaxy and dominate the structure. And in this Hubble progression from early E systems to late type spirals, the Seares color difference between central regions and the blue arms becomes even more spectacular. These crucial features are illustrated so well in this Atlas as to make Hubble's extrordinary classification system seem simply self-evident.

The next central development came in the decade between 1935 and 1945 with an increasing understanding of the relation amongst the various components of a typical galaxy (nucleus, central bulge, disk and spiral arms). This developed into Baade's population concept when he discovered that the red central region of the Andromeda Nebula (M31) was composed of stars, the brightest of which were much fainter in absolute magnitude than the luminous blue stars such as those in the giant association NGC 206 in M31, illustrated in this Atlas. Working from his prior knowledge of the stellar content of globular clusters, and the discovery by Hubble and himself of RR Lyrae variable stars in the dwarf elliptical galaxy in Fornax – a galaxy of a new type found by Shapley at Harvard in 1938 – Baade identified the stars he had found in the central region of M31 (and which could therefore be inferred to populate the central bulges and lenses of Sa and Sb systems in general as well as the E and S0 types of galaxies) with the brightest red giant stars in globular clusters. Baade recognized the dichotomy between the red and the blue stars in their different spatial positions within a galaxy and in the large ratio of their intrinsic brightness (the red stars are faint, the blue are bright), and labeled the two Population I (brightest stars blue and bright, spiral regions) and Population II (brightest stars red and faint, central regions).

The concept of different stellar populations has proved to be fundamental because it is now known that a separation into population types is a separation by age. This color-age relation, described so directly by Wray's diagram in his introductory explanations, came to be understood in the mid 1950's when the Hertzsprung-Russell diagram of globular clusters could be explained. In 1942 Schonberg and Chandrasekhar mathematically followed the evolution of a model main sequence star as its hydrogen was converted to helium. Their calculations reached a limiting configuration where the model had to fundamentally change for the star to remain stable. In the early 1950's the problem was carried further by the Bondi's in England and especially by Martin Schwarzschild and his students at Princeton. At the same time, the observers at Mount Wilson and Palomar, working under Baade, found the main sequence of the HR diagram of globular clusters and identified the turn-off with the Schonberg-Chandrasekhar limit and the brighter sequences of the globular cluster diagram with the Schwarzschild-Hoyle theoretical evolution into red giants. The way was then open to age-date the globular clusters and, by the same method of identification of the main sequence turn-off, the ages of the younger open clusters were also found, giving a natural explanation of Baade's population types as age groups.

Armed with this knowledge of colors, ages and the color differences at different positions within a galaxy, this Atlas, for the first time, makes available the 'museum collection of bright galaxies' from which a further advance in the understanding of galaxy evolution will be made. From this book, galaxies will be chosen for work with Space Telescope on the resolved stars for distance determination, applicable to the yet unsolved problem of the value of the Hubble constant; for work with ground-based telescopes on star formation rates in galaxies of different Hubble types; for studies of dust distribution; for correlations of disk surface-brightness with, say, the total hydrogen mass; and for a host of other problems. It is for these reasons that this Atlas is so important, and why it will be viewed some years hence as one of the major reference works produced in astronomy in the last half of the 20th century.

Allan Sandage

Introduction

This atlas comprises the color images of more than six hundred galaxies representing essentially all recognized form and luminosity classes as well as many peculiar and interacting systems, both within the Local Supercluster of galaxies and beyond. These color images offer a source of aesthetic appreciation and scientific investigation, both of which are fully intended and encouraged. Unlike ordinary color photography, however, the overall method by which these images were obtained and produced was developed specifically to assure their scientific applicability.

Thus the color images in this atlas illustrate the presence and extent within these galaxies of both the very young populations of stars characterized by their 'blue-knot'-like appearance and the very old populations characterized by their yellow color and relatively smooth distribution. Between these two extremes there occur many intermediate stages of color, the interpretation of which in terms of intermediate age and evolving chemical abundances is not always certain.

Yet such is the purpose of this Color Atlas of Galaxies, to serve as a basis for expanding our insight into the nature of galaxies, to raise more issues than it resolves and thus to stimulate inquiry into the detailed nature of the evolutionary processes which are at work gradually transforming galaxies from youthful clouds of proto-galaxian gas to old, declining systems and beyond.

This work then is intended to provide you with an opportunity to discover and examine some of these processes yourself, to question the possibility of their reality and to validate or repudiate these tentative hypotheses through your own better observations, theoretical considerations, or understanding of their nature.

Observational Selection

The galaxies initially selected for observation were chosen on the basis of size from among the NGC objects listed in the Reference Catalog of Bright Galaxies (RC1). With the exception of NGC 224 (M 31), NGC 598 (M 33) and the Magellanic Clouds, together with several objects with very low surface brightness, the atlas is virtually complete with respect to galaxies with $\log D(0) \geqslant 1.50$ (3.2 arc min.) comprising 222 objects. For $1.50 > \log D(0) \geqslant 1.40$ (2.5 arc min.) the atlas contains 127 galaxies, 81% of those listed in the RC1.

The 259 remaining objects listed in the Table of galaxies illustrated in the atlas (see page 189) are either not among the NGC galaxies in the RC1 or are smaller than $\log D(0) = 1.40$. These include galaxies in several fields in the Hercules supercluster, as well as other close groups and interacting systems of galaxies. Many of the smaller galaxies were photographed with the larger scale telescopes, but because large telescope time is at a tremendous premium additional selection criteria were incorporated. These included: galaxies with known or suspected nuclear activity; S0 systems with dust which the author suspected of possibly containing regions of star formation activity (disallowed by S0 classification criteria), e.g. NGC 4710; a sample of elliptical galaxies (underrepresented by the $\log D(0)$ criteria); close pairs of galaxies; and distorted or otherwise peculiar systems.

All of the observations in the northern hemisphere were obtained by the author at the McDonald Observatory of the University of Texas. Many southern galaxies are inaccessible from that latitude, however, and the problem is compounded by the specification that air mass not exceed 1.5 as a practical working limit in order to avoid significant color artifacts due to differential extinction.

In order to achieve a reasonable degree of completeness it was also necessary to obtain observations in the southern hemisphere. To this end both the Las Campanas Observatory and the Cerro Tololo Inter-American Observatory made telescope time available to the author with the result that both the northern and southern hemispheres are equally represented with respect to galaxies selected under the criterion $\log D(0) > 1.4$. Furthermore, for smaller RC1–NGC galaxies with $1.40 \geqslant \log D(0) > 1.30$ (2.0 arc min.) the atlas contains 48 of 102 such southern objects (47%), while for the northern objects the atlas contains 62 of 141 (44%) of the candidates in this size range. As a result of these efforts the atlas is reasonably representative of the entire sky.

Technical Aspects

The detector-filter selection was based on both scientific and technical considerations. The overriding consideration was the need for the colors to be reliably interpretable. In order to achieve this the colors had to be both technically accurate and scientifically significant. Ordinary emulsions used in color photography were less than satisfactory in both regards.

In the technical sense color photographic emulsions suffer from what is termed 'color tracking' errors, the failure to reproduce a given input color with a corresponding single output color independent of intensity. The result off such an error is that the color seen in the image represents in part the brightness of an object instead of its color. A number of photographs of astronomical objects have been published which exhibit this flaw and are consequently difficult if not impossible to interpret correctly.

Accurate color tracking requires that the image brightness correspond to the object intensity in precisely the same way for each of the three color passbands in the system. In photographic terms the gammas must be identical for all three colors. Since gamma varies with wave length for photographic emulsions, any system incorporating purely photographic processes cannot avoid having to deal with this problem. An image tube, on the other hand, serves not only as an image intensifier and a high quantum efficiency detector, but most importantly in a color application it serves as an image converter, changing whatever the input color passband to a single output passband of the color of the phosphor glow. In the case of the RCA-Carnegie two-stage image tube selected for this program the output phosphor image is blue. The three blue-light images representing the three separate input colors are recorded on Kodak IIa–O emulsion and processed simultaneously to yield virtually identical gammas for the three color passbands. Although other small effects do enter in to down-grade the results slightly from theoretical perfection, the practical result nevertheless is the achievement of excellent color tracking as evidenced by the color photographs themselves. Note in this regard that in the images of bright stars the color of a star is maintained throughout its (stray light) image regardless of brightness. See for example the NGC 45, NGC 278 and NGC 4217 fields.

The requirement that the images be both reliably interpretable and scientifically relevant was a major factor in the selection of appropriate passbands. Five principal considerations led to the adoption of passbands approximating as closely as possible the UBV photoelectric system. (i) The UBV system, being biased towards the ultraviolet from the normal blue-green-red of standard color systems, favors the detection of hot stars which represent the young stellar population and hence serve as an indicator of present star formation activity in galaxies. (ii) The passbands of the UBV system are sensitive primarily to the light arising from stellar radiation. Only about 10% or less of the light from a typical HII region transmitted by these passbands is from the gaseous (non-stellar) component while 90% or more is from the stellar component. Hence interpretation in terms of stellar content can be made with considerably more certainty than is possible with a system which detects H-alpha emission of ionized hydrogen for example. (iii) A large amount of photoelectric data in the UBV system already exists for galaxies; hence the interpretation of the color perceived in the color images can be facilitated by comparison with these data. Accordingly UBV data accompany each galaxy in the atlas when available. (iv) The effects of evolution on the brightness and color of individual stars based on both observation and theory are already well established in the scientific literature, either directly in the UBV system or readily adaptable to it. Thus there is a basis for interpretation of the UBV colors of stellar systems in terms of the evolutionary characteristics of their individual stars. (v) The UBV system with its relatively broad passbands permits shorter exposures than do narrower band systems, thereby allowing a larger number of galaxies to be observed. This is an advantage in an initial survey such as this, although in subsequent work of this nature narrow band systems could be selected to facilitate specific studies of particular galaxies.

As a further aid to the interpretation of the color images and estimation of their reliability, a color referent can be established in terms of photometric data for individual bright stars appearing on the color images. For this purpose Harold Corwin has very kindly obtained UBV photometry of the brightest stars appearing in this atlas. These data, together with further discussion, are given in the atlas text beginning with NGC 45.

The observing system parameters are the following:
Filters: U (2.0mm UG2 + 10mm CuSO$_4$); B (2.0mm GG13 + 1.0mm BG12 + 10mm CuSO$_4$); V (2.0mm GG495 + 10mm CuSO$_4$).
Detector: RCA–Carnegie two-stage image tube with S–20 photocathode, quartz faceplate and blue phosphor output.
Transfer Optics: f/1.0 stopped to f/1.4; 1:1 scale.
Photographic Emulsion: Eastman Kodak IIa-O baked 9 hours at 65 degrees C; developed 5 minutes at 20 degrees C in D–19.
Exposure Times: standard; 6 minutes in U, 2 minutes in B and 4 minutes in V at f/13.8, and half these values at f/9.0.
Telescopes and Atlas Print Scales: Cerro Tololo Inter-American Observatory 0.9 meter aperture, f/13.0, 4.45 arc seconds per mm.
Las Campanas Observatory 1.0 meter aperture, f/15.0, 3.80 arc seconds per mm.
(1.85 arc sec/mm full page) (5.20 arc sec/mm NGC 253).
McDonald Observatory 2.7 meter aperture, f/9.0, 2.23 arc seconds per mm.
McDonald Observatory 2.1 meter aperture, f/13.8, 1.90 arc seconds per mm.
McDonald Observatory 0.9 meter aperture, f/13.5, 4.20 arc seconds per mm.

(2.05 arc sec/mm full page).
McDonald Observatory 0.75 meter aperture, f/13.5, 4.85 arc seconds per mm.

The differences in f-ratio as a function of exposure time produce differences in apparent surface brightness of ± 0.2 magnitudes in the atlas photographs. From a photometric point of view the observations comprise essentially raw data corrected only for differential sky brightness to provide a neutral sky in the color images. All other factors affecting apparent brightness and color remain in the atlas images. Nevertheless this collection of galaxy photographs is unique in its general uniformity in exposure and print contrast, such that differences in apparent surface brightness from one galaxy to another signify actual differences between the galaxies and not simply differences in the exposures or the printing. The degree of repeatability of both the color rendition and the surface brightness of the galaxies as observed with different telescopes is illustrated in a number of dual images throughout the atlas.

The Dye Transfer process of Eastman Kodak was used to make the color prints from the three black and white U,B,V images obtained at the telescope. This system is thoroughly described elsewhere (Kodak Dye Transfer Process, Kodak Publication E–80), but because its little known properties are ideally suited to this application, and because the reliability of the colors generated with it is the keystone of the present work, a brief technical overview is warranted.

The process comprises a highly controllable system for generating three separate negative-color dye images and for depositing these images in juxtaposition within a single layer of emulsion either on a paper base for a reflection color print or on a film base for a color transparency.

The first step consists of generating the 'matrices' which define the individual dye images. In this stage the three images obtained at the telescope serve as the 'separation negatives'. The separation negatives are placed in an enlarger individually and the images are exposed onto the matrix film, using an on-easel photometer to set the sky intensity to the same value for all three exposures. By keeping exposure times equal for the three matrices, differential reciprocity failure is avoided. The identical sky exposures produce identical sky densities on the three matrices. Since this procedure introduces a small color component to the rest of the image, separation negatives with aberrant relative sky densities are rejected and the object is either reobserved or deleted. The matrix film is exposed through the transparent backing, emulsion side down. The matrix film emulsion is sensitive to blue light, and it contains a yellow dye. The last three factors are fundamental to the dye transfer process itself.

The matrix film positives are developed simultaneously to produce identical gammas (necessary for correct color tracking). The matrix development process raises the melting point of the exposed gelatin. After fixing, one more step is necessary to complete the matrix processing: the hot water (59 degrees C) rinse. During the hot water rinse the unexposed gelatin is washed off of the matrix. Since the emulsion is blue sensitive, yet contains a yellow dye, the depth of penetration of the exposure is exponentially dependent on the intensity.

The finished matrix is a three dimensional gelatin sponge, the thickness of which is a highly linear function of log exposure. When this matrix is soaked to saturation in an appropriate dye the amount of dye contained at a given point is physically determined by the thickness of the matrix, and hence the resulting absorption or color density due to that dye in the final image is in exact one-to-one correspondence to the log transmission through the original separation negative.

The dyes (cyan, magenta and yellow) in equal amounts produce a perfectly neutral grey scale. This test is easily performed using a single matrix to transfer the three dyes. Any deviation from balance as measured with a three-color reflection densitometer can be corrected by appropriate additives to the dye bath. The dye process is completely independent of variations in temperature, and is carried out in ordinary room light. The matrices are aligned by means of star images on a light table for register punching. Superposition of the three dye images on the print during the dye transfer phase is by means of register pins on the transfer board.

Small corrections to color or overall density (all three colors) are possible by soaking the matrices in weak dye baths. Repeated transfers can be made with each of the colors to increase density.

For this program the matrices were exposed to a sky density suited to a triple transfer (each of the colors applied three times) to optimize final print contrast and sky density, and to provide at the same time a matrix well suited for making small correction transfers. The sky density and neutrality (same density in all three colors) were measured at two places on each final print. Correction transfers were applied if necessary. The mean sky density was 2.0±0.10s.d., while for individual colors the width of the maximum envelope for the three separate densities was .07d.u. Thus there is a very sound basis for the validity of these color images.

Color Properties of Galaxies

The stellar component of galaxies is primarily responsible for their color in the visible and near-visible ultraviolet region of the spectrum. Two other principal components, gas and dust, also influence the color of galaxies, but to a much smaller degree than the stars themselves.

The color of an individual star is determined primarily by the temperature at its surface, ranging from blue for extremely hot stars to red for very cool stars. The color is also influenced to a lesser degree by selective absorption attributable to cooler gases in the star's atmosphere. The temperature at the star's surface depends on its size and total luminosity. The luminosity depends largely on the mass of the star while the size depends to a great extent on internal conditions within the star.

As a star ages these internal conditions change markedly while the mass changes only slightly. The result for a typical star is that after spending the major part of its life at its original brightness and color it suddenly begins to expand and cool, becoming increasingly redder. During this phase the star becomes more susceptible to mass loss, losing mass not only by radiation but also increasingly by streaming of atoms, molecules and even minute dust grains directly from its vastly distended outer atmosphere. Eventually the star loses its entire outer shell which dissipates into the interstellar medium. If the star is excessively massive it simply explodes during this phase, becoming a supernova. A recent supernova in NGC 4321 is illustrated in the atlas.

Thus a star, born within a cloud of gas and dust, gradually evolves and dies, returning a significant portion of its original constituents back into the interstellar medium from whence it came. Because the nuclear reactions going on within the star throughout its life have steadily converted lighter elements to heavier ones the gas recycled from the star back into the interstellar medium is richer in heavier elements than it was originally.

Thus the interstellar medium also evolves. What was once a mixture of only hydrogen and helium develops an abundance of heavier elements, and the properties of the stars formed from the evolving mixture of gas and dust are changed accordingly. Whereas the earliest stars formed in a galaxy are very metal-poor, stars formed later in the same galaxy are metal-rich. The principal effect on the colors of stars and stellar systems is that metal-poor red giant stars are measurably bluer than those with 'normal' abundances.

Research on stellar evolution has progressed a great deal during the past three decades as revealed by the remarkably detailed evolutionary models for stars of widely differing mass and metallicity now available in the literature. The evolutionary nature of galaxies, however, still remains largely conceptual, and tentative in its details. By applying detailed knowledge of stellar evolution we can proceed to develop more advanced models for the evolution of the aggregate color of a system of stars as it ages.

A simple model of an evolving stellar system is presented here to provide further insight into interpretation of the color images in the atlas. In figure 1 approximate evolutionary tracks are indicated for stars which begin their lives on the main sequence with masses ranging from 40 solar masses to 0.2 solar masses. The adopted positions of these stars in the color magnitude diagram (M_V versus B−V) for seven epochs (ranging from zero age (0), to thirty billion years (6)) are shown as open circles with the accompanying number(s) of the epoch(s) at that point. In the case of stars on the giant branches the numbers in parentheses represent the adopted fraction of the epoch spent there. By convolving the evolutionary tracks for these stars (in this case a very simple seven mass seven epoch model) with a mass birth function (the number of stars in each mass category born per unit volume of space in a single burst of star formation) the luminosity and color of the resulting system can be computed as a function of age. Note that the most massive stars evolve rapidly and disappear, while the less massive stars systematically become redder (except for the faintest stars which slowly become brighter and slightly bluer). The net effect is that from the moment star formation ceases the system becomes progressively redder.

Table I lists data on the seven mass categories used in the model together with three different mass birth functions, the first of which is an approximation of the present mass function for stars in the solar neighborhood. The other two mass birth functions represent extremes beyond reasonable expectation for excess high mass stars in the first case and excess low mass stars in the other.

The results of the calculation are listed in Table II. Current models for star formation indicate that for the first several epochs the newly born stars would still be hidden by the dust cloud in which they formed. Thus these results are in reasonable accord with photoelectric measurements of the colors of individual blue knots in galaxies which have values typically of B−V=0.2. For systems with pure old yellow stellar populations the photoelectric measurements are typically on the order of B−V=0.95. This simple model is in excellent agreement with these colors assuming system ages on the order of 10^{10} years.

Actually these results are largely fortuitous, as this model is meant primarily to serve as an illustration of the concept, and a much more detailed one would be required to produce results that could be considered significant. Nevertheless, the trend in color with age is clearly from blue to red, with the blue of objects like the blue knots in NGC 1042 corresponding to an age on the order of 10^7 years (i.e. star formation is still taking place), while the reddish yellow color of NGC 936 (a pure old yellow population) corresponds to an age on the order of 10^{10} years; and even this simple model is sufficient to illustrate that point.

Finally, these results assume normal metal abundances in the stars. For metal-poor stars the red giant branch is shifted as much as 0.25 magnitudes to the blue. Therefore it is possible that old metal-poor stellar systems could have colors as blue as B−V=0.7.

Dust and gas can each influence the observed color of a galaxy directly.
The effect of gas, such as a bright HII region, is generally insignificant in these photographs, however, because (as explained previously) the light from the exciting stars is dominant in the UBV pass-bands. Only in instances where the object is near enough for the HII region to be resolved is the gas detectable with a characteristic violet color. See for example NGC 604 and the discussion of NGC 55.

The effect of dust, however, can be extreme, in both its reddening and dimming effects on light passing through it. See for example NGC 2146. Galaxies seen at low galactic latitudes (near the plane of the Milky Way) are appreciably reddened by dust in our own galaxy. In additon to the optical effects of dust, the material plays a definite role in the evolutionary processes in galaxies, but the full extent of its role may have yet to be acknowledged. These photographs may help to reveal some of the more dynamic roles which dust (in association with gas) could play in the synergetics of galactic evolution.

In conclusion this color atlas of galaxies offers a number of opportunities beyond the obvious testing of relatively simple hypotheses by inspection. It offers an opportunity to analyze galaxies in terms of localized regions of distinct population types and to recognize numerous regions of similar nature suitable for more detailed investigation. This in turn supports the development of more elaborate and more comprehensive models, together with the concepts for testing them.

Finally, the atlas offers us an opportunity to relax and examine these immense assemblages of stars and gas and dust together with occasional planetary systems, with the view of a naturalist; to broaden our perspectives and insight into the nature of galaxies and the processes which drive them from birth to birth through eternity.

Table I
Sample mass birth functions

Case M/M$_\odot$	I log N	II log N	III log N
40	4	7	3
7	6	7	4
2	6.5	7	5
1.4	7	7	6
0.9	7.5	7	7
0.6	8	7	8
0.2	8.5	7	9

Table II
System luminosity and color with age

T	log age	Case I		Case II		Case III	
		M$_v$	B–V	M$_v$	B–V	M$_v$	B–V
0	5.0	−17.4	−.26	−23.8	−.50	−14.5	−.13
1	6.0	−17.4	−.26	−23.8	−.50	−14.5	−.13
2	7.0	−17.0	−.02	−22.1	0.62	−13.9	0.59
3	8.0	−17.2	0.54	−19.6	0.56	−13.7	0.73
4	9.0	−15.4	0.69	−16.2	0.74	−13.5	0.87
5	10.0	−14.6	0.96	−14.2	0.97	−13.3	1.01
6	10.5	−15.6	1.10	−13.9	1.10	−15.3	1.11

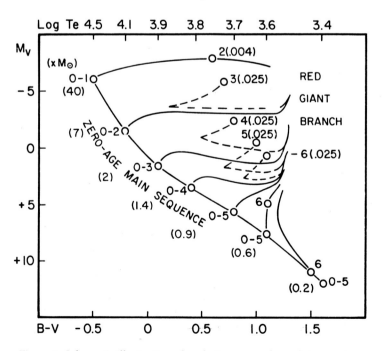

Figure 1. Schematic illustration of evolutionary tracks and time scales for a representative range of stellar masses. Refer to text for further description.

Supplementary Data

A supplementary data table accompanies each atlas illustration. The first line is the NGC number of the galaxy illustrated. If more than one galaxy appears in the field the NGC number of one of the galaxies in the field is selected arbitrarily to identify the field. In any case the data on the galaxy given in the data block pertains to the galaxy named at the head of the block. Identifications of other galaxies appearing in the field are generally given in the descriptive text.

In the following refer to NGC 0024 (atlas page 1) as the example.

The second line gives the observing data: Plate Number (1392); exposure duration for the U plate in minutes (6); exposure date listed as year, month, day and hour to the nearest tenth for the mid point of the exposure sequence, in Universal Time (81 11 01 03.4). One exception is that the date for the 900b plate series is given to the nearest tenth of a month.

The last entry on the second line gives the observatory and telescope with which the observation was made. Observations from the northern hemisphere were made at the McDonald Observatory (McD), while those from the southern hemisphere were made at the Las Campanas Observatory (LC) and the Cerro Tololo Inter-American Observatory (CTIO). Telescopes are identified by their aperture to the nearest tenth of a meter (1.0m). Refer to the table of plate scales given in the section on observing system parameters (pp vii–viii).

The third line lists available photometric data, generally from the RC2 augmented by unpublished data kindly provided by H.Corwin in advance of publication (to be published in the RC3). In instances where this data remains approximate the values are given to the nearest tenth of a magnitude. Magnitudes are total magnitudes uncorrected for galactic extinction. Because these measurements are sensitive to about two magnitudes fainter in surface brightness than the Atlas photographs the photoelectric colors can be biased towards the color of the faint outer region in a direct comparison with the perceived Atlas colors, particularly for low surface brightness galaxies. Nevertheless this comparison will be found useful and informative. In aggregate groups of stars the $B-V$ and $U-B$ color indices are strongly correlated, hence in most instances the $B-V$ index alone serves as a useful indicator. For further discussion refer to the section on interpretation of the Atlas photographs. The last entry on this line is the radial velocity corrected for galactic rotation.

The fourth line gives classification data on the system of de Vaucouleurs (.SAS5..) as given in the RC2 and on the system of Hubble and Sandage (Sc(s)II–III) as given in the RSA. The next two entries are the Type and Luminosity Class values on the system of de Vaucouleurs. The last entry is the direction of north in the field in units of ten degrees measured from zero at the top of the illustration clockwise. The fields appear as projected on the sky, hence for the example north is to the right and east is to the top.

Positional notation in the descriptive text is given in a radial coordinate system with the angle designated as described in the preceding paragraph and the length of the radius vector in units of one tenth of the field radius. At the atlas print scale the radial unit corresponds to approximately one centimeter on all but full page illustrations. For the latter the radial scale is approximately 1.6 cm per unit (except for NGC 253 where the value is approximately .8 cm per unit). The scale of the atlas prints in arc seconds per millimeter is given in the section on observing system parameters.

NGC 0013

929 3 79 11 25 3.8			McD 2.7m
.RS..1*.		T=1	N18

This galaxy exhibits a rather uniform disk which, although poorly resolved, reveals evidence of star formation regions both near its outer boundary at the upper left and in an inner ring surrounding the nuclear region.

NGC 0023

990 3 79 11 26 04.0			McD 2.7m
B=12.8	B−V=0.81		V=4793
.SBS1..	SbI−II	T=1	N27

A rectilinear array of star-formation regions ('blue knots') precedes a massive dust lane. Several apparent blue knots are just detectable at the edge of the bright nuclear region.

NGC 0024

1392 6 81 11 01 03.4			LC 1.0m
B=12.1	B−V=0.60	U−B=−.04	V=597
.SAS5..	Sc(s)II−III	T=5 L=5	N09

Star formation regions are seen scattered throughout a relatively uniform disk. While spiral nature is evident in many small spiral features, large scale spiral arm structure is only weakly manifested. It seems likely that much of the star formation activity seen here is due to stochastic processes. A weak nuclear region is discernable by its characteristic yellow color. This system is 0.2 mag. bluer than the preceding galaxy NGC 0023.

NGC 0045

1452 6 81 12 18 02.5			McD 0.7m
B=11.1	B−V=0.69	U−B=−.04	V=508
.SAS8..	Scd(s)III	T=8 L=8	N07

This is one of the lowest surface brightness galaxies in the atlas. Several faint blue knots are visible on the original print. Harold Corwin has obtained photoelectric photometry of the bright orange star in this field (as well as photometry for 54 other stars illustrated in the atlas). His measurements, published here for the first time, allow you to make your own assessment of the utility of these color images as semi-quantitative data, and to establish your own relation between the quantitative photoelectric data and the apparent color as perceived by you individually. You will note that the perceived color of stars differs rather systematically from the perceived color of galaxies of the same U−B and B−V, the galaxies appearing about 0.1 to 0.2 mag. redder. This seems to be understood in terms of differences between the overall energy distribution for stars and for systems of stars. For this star V=6.85, B−V=1.11, U−B=1.00.

NGC 0055

1319 6 81 10 29 04.7			LC 1.0m
B=7.9	B−V=0.50	U−B=−.20	V=98
.SBS9*/	Sc	T=9	N05

Only the central region of this galaxy is included in the field. Compare the color with NGC 0024 above. At B−V=0.50, this is a relatively blue galaxy. Note the blue knots at the upper left (33,4) which are resolved into hot supergiants, as contrasted with the violet appearing knots grouped near the center of the field. These latter provide an excellent example of OB associations resolving into HII regions. When the extent of the gas is resolved the color of the emission region itself is perceived. Since the red H alpha line is rejected by the UBV passbands, the dominant lines become those of singly and doubly ionized oxygen in the U and V passbands respectively. Since these record as blue and red respectively, the perceived color is violet. Variations in the relative strengths of the lines affect the color accordingly. Of course other lines contribute to the color as well, but this is the essential basis for understanding the appearance of HII regions in these photographs. Finally, spectrophotometry of several HII regions in other galaxies by Marshall McCall indicates that typically the ratio of flux from stars to flux from gas is about 10:1 or greater. Thus in cases where the gas is not resolved the light from the hot blue stars dominates the perceived color, hence the blue knots. Generally, then, the colors seen in these photographs may be interpreted unequivocally in terms of the stellar content, together, of course, with the effects of scattering and extinction by dust. In this galaxy the apparent presence of dust (seen by extinction) is not overwhelmingly confirmed by reddening, and at least some of the irregularity of light distribution seen here may be due to actual irregularities in the surface distribuion of stars, rather than dust as it might appear.

Individual supergiant stars are seen to be resolved in this galaxy at an apparent magnitude of approximately +20. The corresponding distance modulus, by visual inspection, is approximately +27 or +28. Refer to the discussion accompanying NGC 300, NGC 604 and NGC 625.

NGC 0068

930 3 79 11 25 04.0			McD 2.7m
B=13.9	B−V=0.99		V=5941
.LA.−..		T=3	N05

NGC 68 is at the lower right (12,4). Also included are the spiral NGC 70 at the upper right (07,3), NGC 71 at the center, and the barred spiral NGC 72 at the left (27,6). The blue spiral pattern of NGC 70 is just discernible. A faint blue stellar object is visible at 16,3, and a faint blue galaxy is visible at 31,3.

NGC 0095

991 3 79 11 26 04.1			McD 2.7m
B=13.2	B−V=0.70	U−B=0.07	V=5064
.SXT5P.	Scd(s)III	T=5	N00

The very faint overall spiral pattern appears as numerous spiral features of relatively high angle of incidence, with enhancements which as a group appear as a major spiral arm.

NGC 0100

0100
0128
0127
0130
0134
0147
0150
0151

```
931  3  79 11 25 04.1              McD 2.7m
B=13.9    B−V=0.75   U−B=0.00    V=1027
.S..6*/                    T=6           N02
```

Although this system lacks any massive dust lane, it appears to have a slight overall reddening probably due to dust. The highest surface brightness features are regions of star formation. There is no evident change in color towards the nuclear region. Note the extremely red distant galaxy in the lower right corner (11,6). The color of this galaxy could easily correspond to a k-correction of 0.3.

NGC 0134

```
1341  6  81 10 30 03.0             LC 1.0m
B=11.0    B−V=0.88   U−B=0.29    V=1531
.SXS4..   Sbc(s)II−III      T=4           N32
```

Note the widespread occurrence of sharply concentrated regions of star formation, the blue knots. These regions extend well into the central area of the galaxy which is dominated by a relatively smooth older (yellow) population of stars.

NGC 0147

```
1453  6  81 12 18 03.0             McD 0.7m
B=10.4    B−V=0.94   U−B=0.25    V=−11
.E.5.P.   dE5              T=5           N00
```

The surface brightness of this galaxy is just above the detection limit of the system. A number of stellar appearing objects of marginally blue color are seen in the field.

NGC 0150

```
1430  6  81 11 03 05.3             LC 1.0m
B=11.7    B−V=0.62   U−B=−.02    V=1537
.SBT3*.   Sbc(s)II         T=3 L=1*  N18
```

The spiral region is dominated by active star formation. Note the sharply resolved blue knots along the lower arm. The small, sharply defined nuclear region exhibits the yellow color characteristic of an old stellar population.

NGC 0151

```
411  6  77 11 06 04.8              McD 2.1m
B=12.2    B−V=0.74   U−B=0.13    V=3746
.SBR4..   SBbc(rs)II       T=4 L=3*  N16
```

Note that the nuclear region is divided into three parts, a small round nucleus flanked by two crescent shaped arcs at 90 degrees to the bar axis. The bar consists of two distinct regions of similarly old stars, one of which is sharply bifurcated, probably by dust. These two regions terminate in what appear to be dust lanes along the inside edge of a ring of active star formation regions. Tracing this ring structure counter-clockwise reveals that each of the two ring segments originates at the trailing edge of the two 'yellow knots' which define the bar. Note that the yellow color of the bar extends into the ring on the upper right, despite the fact that this ring is generally dominated by star formation. Arguing that discrete structure seen in yellow light, and hence composed of old stars, must, therefore, be a 'stable' or, at most, a very slowly evolving configuration in surface brightness structure, the juxtaposition of the bar and ring structures is interesting. If ring and bar co-rotate at these points, why has the star formation rate been different at different places in the ring during the past 10^9 years or so? If they do not co-rotate then stars of the bar system would sweep through the same space occupied by stars of the ring system traveling at a different velocity.

Several background galaxies are visible. Two are near the edge of the field at the upper left, one blue and one red. A third galaxy is seen as a diffuse yellow knot superimposed almost exactly on the spiral arm of NGC 151 at the lower right (11,5). It is above and slightly to the left of the bright foreground star which appears projected on the end of the same spiral arm feature. If this third galaxy is a background object then it could provide a means of studying the interstellar medium in NGC 151.

NGC 0128

```
342  6  77 11 05 04.3              McD 2.1m
B=12.6    B−V=1.00   U−B=0.59    V=4384
.L...P/ SO2(8)p            T=2           N05
```

The color of NGC 128 appears to be quite uniform, even in the dark areas cutting the narrow extensions. The apparent structure certainly does not seem to be greatly influenced by dust. It should be possible, therefore, to closely match the brightness distribution of this galaxy with a purely dynamical model. The galaxy to the right of NGC 128 (10,2) is NGC 127. Despite the relative smoothness of NGC 127 the blue color probably indicates the presence of active star formation regions. The galaxy at 32,3 is NGC 130.

NGC 0157

345 6 77 11 05 04.9   McD 2.1m
B=11.0   B−V=0.61   U−B=−.02   V=1749
.SXT4..   Sc(s)I−II   T=4 L=1 N27

This galaxy seems to be dominated by the light of an intermediate age population. The nuclear region is relatively small, and the old (yellow) disk population does not appear to extend significantly beyond the small central region. Several major spiral features are delineated by sharply defined blue knots, active star-formation regions with massive young stars still remaining on the main sequence. Most of the light, however, seems to be coming from neither of these two sources, but instead arises from greenish appearing, almost amorphous regions which blanket the disk of the galaxy. From this it appears that this galaxy has recently (from say 10^8 to 10^7 years ago) undergone a massive, galaxy-wide burst of star formation which is only now tapering off to a more 'normal' rate.

An asteroid is seen at the upper right (06,6) as a red image above a green image. The red image corresponds to V and the green image to B. A very weak blue appearing image, corresponding to the U passband lies between the green and red images. The sequence of exposures was B (2 min. began 04 46 45), U (6 min. began 04 51 00) and V (4 min. began 04 58 00).

NGC 0160

932 3 79 11 25 04.3   McD 2.7m
B=13.4   B−V=0.97   V=5460
RLA.+P.   T=1 N00

This photograph barely detects a faint blue outer ring. Individual blue knots are not detected.

NGC 0185

1454 6 81 12 18 03.3   McD 0.7m
B=10.0   B−V=0.90   U−B=0.30   V=4
.E.3.P.   dE3p   T=5 N00

Although dominated by the old yellow population of stars, this galaxy is noted for its several dust regions visible here, and the presence of blue stars, several of which may be just discernible on this small scale photograph. Reddening due to the dust is evident around the dust patches, particularly the one immediately above right of the nucleus.

NGC 0205

249b 6 77 10 15 07.2   McD 0.9m
B=8.8   B−V=0.84   U−B=0.22   V=1
.E.5.P.   S0/E5p   T=5 N28

Morphologically similar to NGC 185, this galaxy, dominated by the old yellow population of stars also contains dust and a few blue stars characteristic of a young stellar population. Although seemingly enigmatic, this combination occurs frequently and needs to be considered as a norm rather than an unusual chance occurrence in galaxies.
The red streak is a flare from a nearby bright star.

NGC 0210

252b 6 77 10 15 07.8   McD 0.9m
B=11.6   B−V=0.72   U−B=0.05   V=1700
.SXS3..   Sb(rs)I   T=3 L=1 N30

Some blue regions are present in the bright inner disk. Several sharp blue knots are seen in the spiral arms. For comparison, the slightly extended blue region at 28,2 is probably similar to NGC 206. Note also the red background galaxy at 05,2.

NGC 0206 (in M 31)

933 3 79 11 25 04.5   McD 2.7m
B=10.7   B−V=0.71   U−B=−.01   N10

NGC 206 is a resolved OB association in M 31. This photograph shows clearly the nature of a 'blue knot' fully resolved into stars. Blue (and therefore hot, massive and young) stars dominate the field and the composite color of the assemblage. A few red supergiants and fainter red giants are visible, but have only a negligible effect on the composite color of the association. The color of an association like this will not be radically altered until most of the blue stars have evolved, and the brightest stars on the main sequence have become those of spectral class A (which appear blue-green in the color system used in this atlas) or spectral class F (green). As evolution proceeds the overall color becomes influenced by the increasing number of red giants, so the composite color shifts its representation from purely main-sequence for young OB associations to purely giant branch for very old systems whose main-sequence has burned down to stars of very low luminosity.

NGC 0221   M 32

1455 6 81 12 18 03.7   McD 0.7m
B=9.1   B−V=0.94   U−B=0.47   V=21
CE.2...   E2   T=6 N00

An elliptical galaxy dominated by an old (yellow) population of stars. The dark patch at 00,1 is a defect which is present on all plates taken on the 0.7m telescope in December 1981. This defect is also visible on NGC 185 (above) at 33,1.

1394   6   81 11 01 04.0   LC 1.0m
B=8.0   B−V=0.85   U−B=0.38   V=259
.SXS5.. Sc(s)   T=5   N20

This print is a composite of three fields. The
color of the central region reveals the presence
of a large bar structure. A bright but poorly
defined ring structure surrounds the bar. This
inner region contains a great deal of dust in
large patches which redden the object
considerably. Broad, rather diffuse spiral
features dominate the outer disk. These arms
contain a large percentage of intermediate age
population. Blue knots, indicative of current
star formation activity, are prominent in a
small region of the farther arm. Fainter blue
knots are widely scattered throughout the
spiral features. The blue knots and dust lanes
are much more organized in the sense of spiral
pattern than are the diffuse patches of
intermediate color and age.
Corwin's photometry of the bright star gives
V=9.28, B−V=0.69 and U−B=.14.

4

NGC 0247

1320 6 81 10 29 05.0			LC 1.0m
B=9.4	B−V=0.53	U−B=−.05	V=180
.SXS7..	Sc(s)III-IV		T=7 L=7 N23

NGC 247 and NGC 300 (below, right) are rather similar in appearance, differing in surface brightness (or size) and in the number and brightness of the blue knots. The disk of NGC 247 is distinctly yellowish and somewhat smooth, indicating the presence of the usual old population. The nucleus is sharply defined, very compact, and almost white in color,

partly because of its moderately high surface brightness. Isolated stochastic star formation processes would seem to dominate in this system.

Both NGC 247 and NGC 300 resolve into stars of similar brightness, implying that the two galaxies are at similar distances, despite their apparent difference in size.

NGC 0278

934 3 79 11 25 04.7			McD 2.7m
B=11.5	B−V=0.66	U−B=−.05	V=884
.SXT3..	Sbc(s)II.2		T=3 N00

The inner disk is dominated by star formation regions of extremely high surface brightness. Outside of this zone the disk is yellow and relatively smooth. This may be the best example of second wave star formation activity illustrated in the atlas. See NGC 4826. This field contains the bluest foreground star illustrated in the atlas for which photoelectric photometry is available in UBV. Corwin's photometry of this star (SAO 036725) gives: V=8.83, B−V=−0.02, U−B=−0.19. Compare with the star in the NGC 45 field.

NGC 0255

438 6 77 11 07 06.0			McD 2.1m
B=12.3	B−V=0.55	U−B=−.27	V=1873
.SXT4..	SBc(s)II.2		T=4 L=3* N09

The bar in this galaxy appears essentially white, yet it is not strongly overexposed. Compare with NGC 151 for example, for which the system B−V is 0.2 mag. greater than than it is for this galaxy. At best the old stellar population of NGC 255 contributes only weakly to the system luminosity and color. A defect is present at 15,1.

NGC 0289

1342 6 81 10 30 03.2			LC 1.0m
B=11.3	B−V=0.72	U−B=0.12	V=1793
.SBT4..	SBbc(rs)I–II		T=4 L=2* N27

Two massive symmetrical dust lanes obscure enough of the central lens to give it the appearance of an orthogonal bar. Two principal spiral arms are dominated by arcs of star-formation regions which originate at the terminus of the two dust lanes.

NGC 0300

1411 6 81 11 02 04.0			LC 1.0m
B=8.7			V= 97
.SAS7..	ScII.8		T=7 L=6 N24

Although a weak and generally disorganized pattern of spiral features is present in the intermediate age population of stars, the young regions of star formation are scattered more or less randomly across the entire disk of this galaxy. This would therefore appear to be an excellent example of stochastic star-formation processes at work. The nucleus is similar to that of NGC 247, although in this case it appears to be yellow.

Note the resolution into individual supergiant stars, identifiable in the UBV color system by their wide range of color at nearly the same brightness. Stellar appearing objects in galaxies too distant for detection of individual supergiant stars invariably appear blue only, and are actually compact blue knots (similar to NGC 604 or smaller). See for example NGC 3344 and also NGC 7793 which illustrates the problem presented by the stellar appearing blue knots very clearly. As the NGC 604 field shows, however, when individual stars are detectable the brightest stars are found to range widely in color while appearing similar in brightness. It is this property of supergiant stars as a group which facilitates their

identification in UBV color images such as these. Given the observed brightness, or apparent magnitude, and known intrinsic brightness, or absolute magnitude, for such stars one can, in principle, determine their distance. The task of accurately determining the distances of galaxies remains one of the most vexing problems in astronomy, and it would be a disservice to imply, even unintentionally, that the problem is amenable to simple solutions. Nevertheless it is possible to present some of the underlying principles associated with determining extragalactic distances, together with some highly simplified examples derivable from the atlas photographs for purposes of illustration (and demonstration of potential usefulness of the multi-color supergiant class identification criterion) only. Refer to NGC 604 and NGC 625 for further discussion of this topic. The estimated B magnitude of the supergiant branch in NGC 300, based on simple visual inspection, is approximately 19, yielding an uncorrected distance modulus (based on the simplified assumptions outlined in the discussion of NGC 625) of +26 or +27.

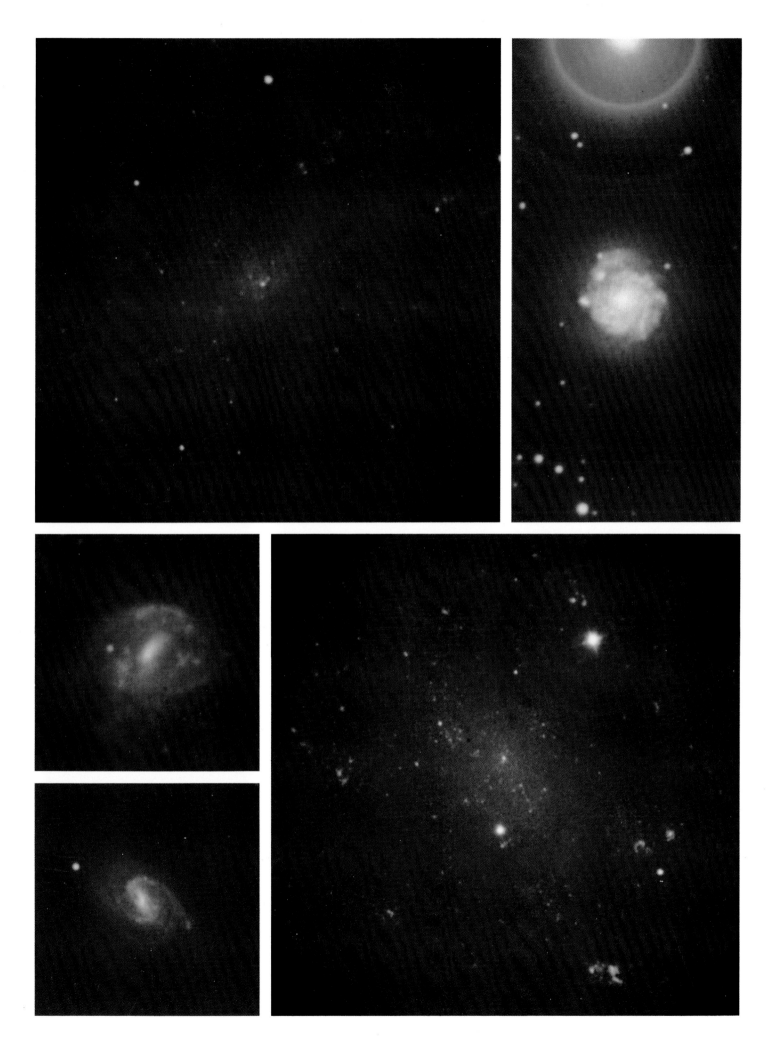

0309
0337
0406
0428
0450
0470
0474

NGC 0309

354 6 77 11 05 06.4   McD 2.1m
B=12.4   B−V=0.61   U−B=0.00   V=5740
.SXR5..   T=5 L=2 N30

The nuclear region gives the impression of a short bar orthogonal to the major axis of the inner ring of young to intermediate age stars. Note the inter-arm plumes extending between the inner ring and the spiral arm structure, particularly in the region 18,1 to 27,2. These plumes are systematically less blue than the blue knots. It is not uncommon in some galaxies to find such plumes apparently attached to blue knots.
A background galaxy is seen at 07,7.

NGC 0337

995 3 79 11 26 04.7   McD 2.7m
B=12.0   B−V=0.45   U−B=−.13   V=1773
.SBS7..   Sc(s)II.2p   T=7 L=4* N13

Perhaps the most remarkable feature of this galaxy is the similarity in structural detail for the yellow and blue regions. Particularly since this structure is relatively patchy. Ordinarily the yellow older population is more smoothly distributed than the young blue population. Dust in the central region may contribute significantly to this appearence. The nucleus appears to be immediately to the left of the blue knot near the center of the galaxy.

NGC 0406

1373 6 81 10 31 04.3   LC 1.0m
B=12.6*   V=1345
.SAS5*.   Sc(s)II   T=5   N18

Several discrete blue knots are seen in this southern galaxy.

NGC 0428

456 6 78 01 08 02.8   McD 0.7m
B=11.8   B−V=0.50   U−B=−.17   V=1266
.SXS9..   Sc(s)III   T=9 L=6 N00

The older population here contributes relatively little to the light and color of this system.

NGC 0450

1456 6 81 12 18 04.1   McD 0.7m
B=12.5   B−V=0.40   U−B=−.17   V=1851
.SXS6*.   Sc(s)II.3   T=6 L=8 N00

Two prominent blue knots punctuate this system. A very red background galaxy is seen at 31,2.

NGC 0470

937 3 79 11 25 05.5   McD 2.7m
B=12.6   B−V=0.69   U−B=0.09   V=2668
.SAT3..   Sbc(s)II.3   T=3   N06

The broad ring appears to be comprised of massive plume-type structures dominated by stars of intermediate age. Several blue knots are visible, mostly in the outer part of the ring.

NGC 0474

420 6 77 11 06 06.5   McD 2.1m
B=12.0   B−V=0.93   U−B=0.35   V=2412
PLAS0..   RS0/a   T=2   N00

Some nonuniform structure is evident in the photograph as a weak arc extending from 28,1 to 35,1.

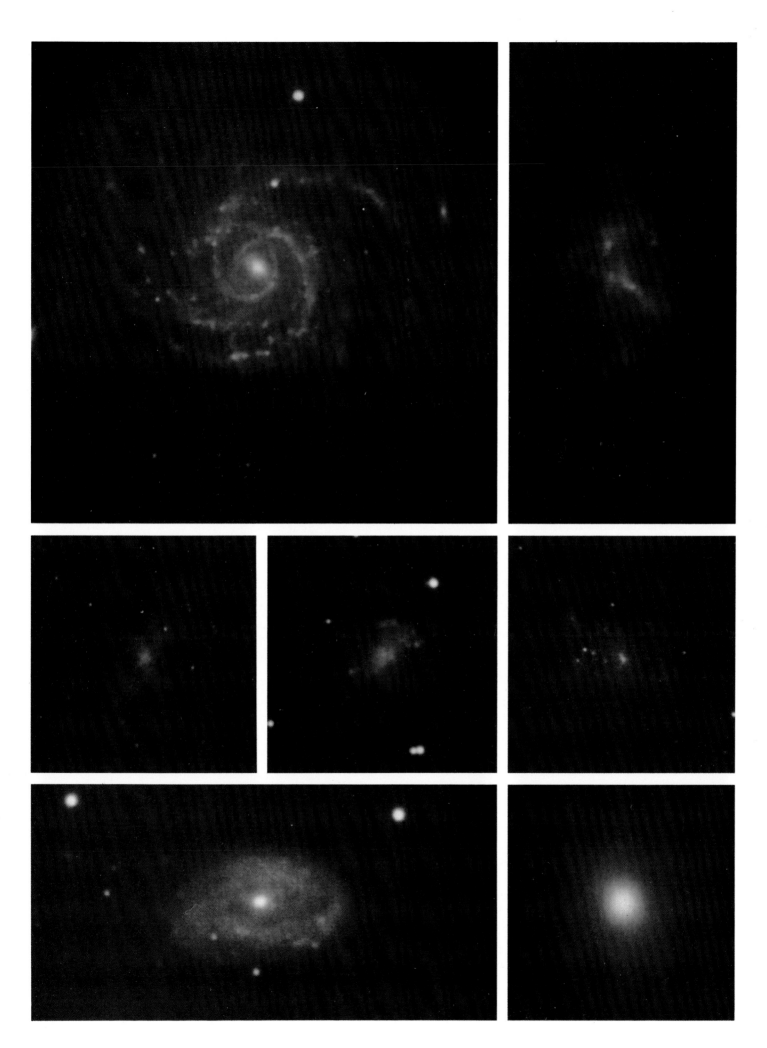

NGC 0488

357 6 77 11 05 06.8   McD 2.1m
B=11.1   B−V=0.86   U−B=0.34   V=2292
SAR3..   Sab(rs)I   T=3 L=1 N27

An incredibly beautiful galaxy. The lens is
surrounded by a rather smooth ring,
apparently cut by dust lanes at several places.
This ring appears yellow and does not seem to
contain any blue knots. The ring is therefore
probably composed almost entirely of an
intermediate to old stellar population. Outside
this ring the narrow winding spiral arms are
delineated by bluish patches and blue knots
characteristic of an intermediate to young
stellar population with a considerable amount
of active star formation. Whatever the arm
formation process may be here, it is very
effective at maintaining the resolution of
closely spaced discrete arms through arcs of
180 to nearly 360 degrees in length.

NGC 0514

360 6 77 11 05 07.2   McD 2.1m
B=12.5   B−V=0.58   V=2615
.SXT5..   Sc(s)II   T=5 L=2* N12

There is an indication here of a short weak bar.
The galaxy is generally dominated by regions
of intermediate age. Star formation is
widespread. Exceedingly long plumes connect
blue knots in the lower inner arm with blue
knots in the lower outer arm. Compare with
the region in NGC 309 at 21,2 in that
illustration.
The red streaks are caused by stray light from a
bright star outside the field.

NGC 0520

363 6 77 11 05 07.6   McD 2.1m
B=12.0   B=V0.85   U−B=0.20   V=2272
.P.....   Amorphous   T=0   N34

The stellar population appears to be mostly
old, based on its relatively smooth distribution
and yellowish color. Bright regions and
embedded knots appear slightly bluish,
however, indicating the presence of young
stars. The system is highly distorted,
apparently having recently undergone a close
encounter with another galaxy. The dust
appears to be oriented systematically with
relation to the stellar component. Note the
reddening due to the dust.

NGC 0521

423 6 77 11 06 06.9   McD 2.1m
B=12.5   B−V=0.80   U−B=0.25   V=5099
.SBR4..   SBc(rs)I   T=4   N00

The well defined bar contains a bright inner
lens which appears very slightly bluish
indicating the possible presence of a small ring-
like structure with active star formation. Just
outside this inner lens, but within the central
bar, a second ring structure can be seen. The
spiral pattern appears to be mostly
intermediate in age, with only a few isolated
blue knots testifying to current star formation.
A group of galaxies is seen on the right side of
the field. One of these galaxies appears very
red, indicative of a background galaxy, while
several of the others appear blue despite
probable Doppler reddening.

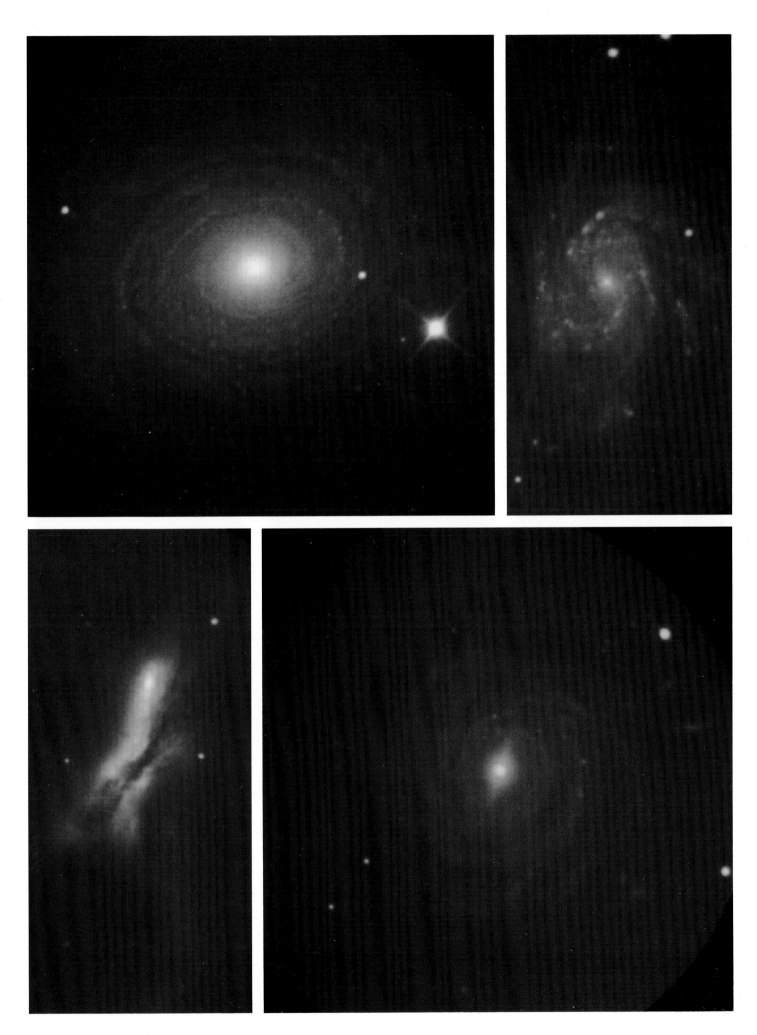

0524
0578
0600
0604
0613
0625

NGC 0524

312 6 77 10 17 08.0   McD 2.1m
B=11.5   B−V=0.95   U−B=0.60   V=2595
.LAT+..   S02/Sa   T=1 N00

As in the case of NGC 474, this galaxy appears to contain several very weak spiral arcs containing several faint bluish appearing knots. One of these knots is at 00,2, and several weak arcs extend from 26,2 to 34,2 and 30,1 to 34,1.

NGC 0600

938 3 79 11 25 05.7   McD 2.7m
B=13.0   B−V=0.59   U−B=−.09   V=1921
.SBT7..   T=7 N04

The bar is divided into three parts, a yellow nuclear region and two outer segments, both of which are green. The internal structure of these regions is not resolved, so it is not clear if the outer bar is a mixture of young (blue) plus old (yellow) giving the appearance of green, or if it is primarily of an intermediate age population (green). The outer structure consists of blue spiral features.

NGC 0613

1431 6 81 11 03 05.6   LC 1.0m
B=10.7   B−V=0.76   U−B=0.15   V=1462
.SBT4..   SBb(rs)II   T=4 L=2 N06

Note the small spiral pattern in the nuclear region which appears to be composed of old stars. Two major dust lanes extend along the bar into the nuclear region and may be the cause of the apparent spiral pattern in that region. The bar seems to have a mixed population of stars. Zones of very active star formation begin at the ends of the bar. The upper arm bifurcates sharply, at nearly a right angle, but only the leading component exhibits strong star formation. In contrast, the trailing arm segment appears to be dominated by stars of intermediate age. The other main arm also bifurcates, but not as dramatically. The angle is about 30 degrees, and there does not appear to be any difference in the populations of the two segments.
The green field star was measured by Corwin: V=9.58, B−V=0.41, U−B=−0.04.

NGC 0578

1343 6 81 10 30 03.6   LC 1.0m
B=11.5   B−V=0.57   U−B=−.14   V=1689
.SXT5..   Sc(s)I–II   T=5 L=3 N18

The short bar-like structure is distinctly yellower than the adjacent spiral features, although it does contain regions of star formation. The disk is dominated by active star formation arrayed in a spiral pattern. Three background galaxies are visible, two of which are strongly reddened. Both reddened galaxies are seen through the disk of NGC 578, and could be bright enough to permit some study of its interstellar medium.

NGC 0604 (in M 33)

936 3 79 11 25 05.3   McD 2.7m
B=12.1   B−V=−.15   U−B=−1.1   V=−52
                                          N00

NGC 604 is the brightest HII region in M 33. This is the type example of a (bright) blue knot, as seen in a nearby galaxy. It easily resolves into individual stars as well as gas. The violet color arises chiefly from the emissions of singly and doubly ionized oxygen in the HII region. (See also commentary on similar regions in NGC 55).
Note the supergiant field stars in M 33. A wide range of color is exhibited in these stars, including green (spectral type F to early G) and yellow (spectral type G). The occurrence of red, yellow and green stars appearing along with blue stellar objects provides a strong indication that the supergiant branch has been detected as individual resolved stars, rather than spuriously as unresolved clusters and close groupings of field stars seen as stellar appearing blue knots. Very few of the galaxies are resolved to this degree in this atlas, but most of them will be so resolved by Space Telescope. In the marginal cases, for very distant galaxies where the determination of accurate distances is critically important, some form of verification of resolution into individual stars will be necessary, and this criterion (wide range of color among (similarly) brightest stellar appearing objects) may be useful.

NGC 0625

1441 6 81 11 04 05.9   LC 1.0m
B=11.7   B−V=0.50   U−B=−.05   V 321
.SBS9$/   Amorphous ImIII   T=9 L=0 N00

There is no clear evidence of a nucleus, or even a nuclear region. Instead, the older yellow population seems to be distributed throughout the galaxy, most strongly influencing the outer region visible here. A number of blue and non-blue stellar objects are resolved in the main body of the galaxy. Based on the discussion of the NGC 604 field and of NGC 300 we may infer that these are supergiant stars in NGC 625. For those who are perhaps not familiar with the various methods used for determining the distances of galaxies we may illustrate here the principle of one such method. Let us suppose for example that the supergiant stars seen here have an apparent V magnitude of +20. Likewise, let us adopt, for this example, an absolute V magnitude for supergiant stars of −7.0, the apparent magnitude they would have if they were only 10 parsecs away (slightly more than the distance to Vega). Then the distance modulus, the dimming due to distance, expressed in magnitudes, is +27.

Since 5 magnitudes corresponds to a factor of 100 in brightness, since magnitudes add logarithmically, and since brightness diminishes as the square of the distance, then for each factor of 100 increase in distance the distance modulus increases by 10 magnitudes. Hence for a distance modulus of 27, the distance must be

$$10 \text{ pc} \times 100 \times 100 \times 10 \times 2.5 = 2.5 \text{ million}$$

parsecs. Of course, a truly definitive distance determination must take into account not only a very carefully measured apparent magnitude, which we do not have here, but also corrections for extinction due to dust and any other factors which could affect the measurements and the assumptions regarding the absolute magnitudes of the objects measured. For example, −8.0 would probably be a better value to use for the absolute magnitude in this case because only the several brightest supergiant stars in the galaxy are visible. In cases such as the NGC 604 field an absolute magnitude of −7.0 is probably more appropriate. Nevertheless, this galaxy serves as an illustration, and an interesting case in point.

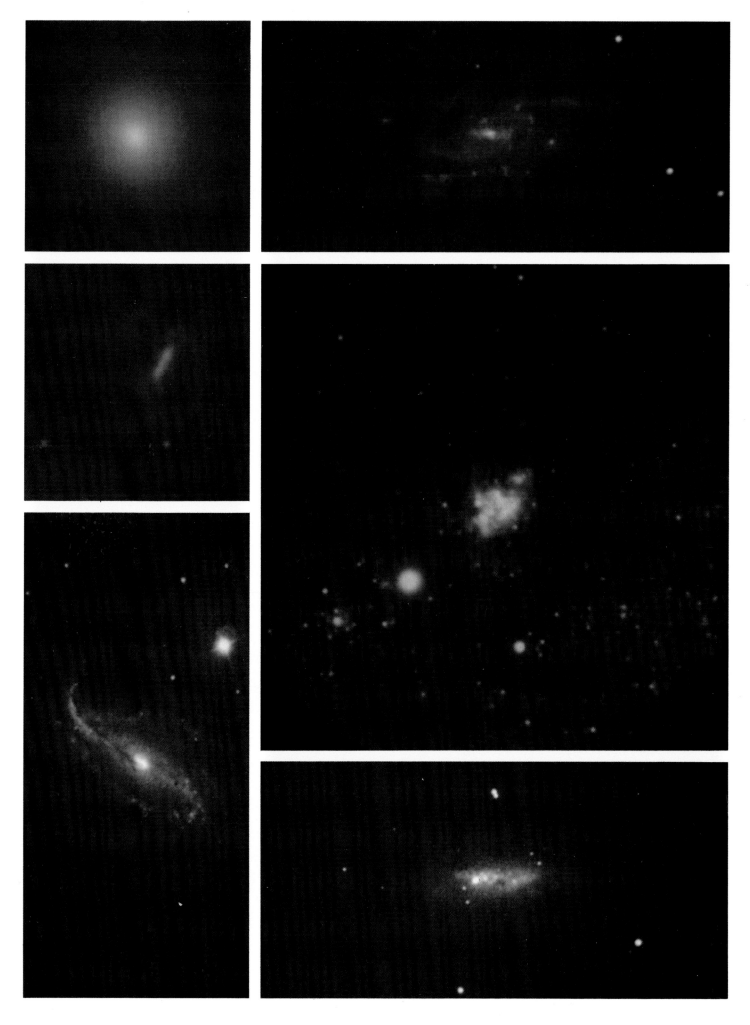

0628
0660
0661
0672
0678
0685

NGC 0628   M 74

258b 6 77 10 15 08.7   McD 0.9m
B=9.7   B−V=0.58   U−B=−.10   V=793
.SAS5..   Sc(s)I   T=5 L=1 N00

The nuclear region is relatively small, but the
old population extends beyond it into the inner
region of spiral structure. Beyond this the color
of the spiral arms suggests a strong
contribution from intermediate age stars. Blue
knots in the inner region are relatively small
and faint. Beyond about 2 units in radius the
blue knots increase sharply in brightness and
number. One expects these parameters to be
directly influenced by the amount of
interstellar gas available for star formation.
Likewise, it can be argued that in a statistical
sense the luminosity of the brightest blue knot
in a galaxy should be correlated with galaxy
type, or, more directly, with total hydrogen
mass.

NGC 0660

939 3 79 11 25 05.9   McD 2.7m
B=11.5   B−V=0.84   V=976
.SBS1P.   T=1   N04

Although the outer boundary of this system is
somewhat smooth or amorphous, the galaxy
exhibits a considerable amount of star
formation activity, evidenced by the blue knots
which are present even in the central region.
The color appears somewhat redder than
might be expected for a galaxy of similar
morphology. The B−V value tends to confirm
this.

NGC 0661

998 3 79 11 26 05.0   McD 2.7m
                                            V=3985
.LA..*.   T=−2   N00

Here we see a system apparently composed
entirely of old stars.

NGC 0672

999 3 79 11 26 05.1   McD 2.7m
B=11.3   B−V=0.59   U−B=−.11   V=578
.SBS6.. SBc(s)III   T=6 L=5 N35

The bar of this galaxy appears to be composed
of a mixture of young, old, and intermediate
age stars. Several compact star fomation
regions appear violet, particularly the one at
27,3. Since this object is nearly stellar the
violet color probably indicates an unusually
luminous HII region. (See NGC 604). The
outer region, although relatively faint, is
dominated by the light of young blue stars.

NGC 0678

997 3 79 11 26 04.9   McD 2.7m
B=13.2   B−V=1.05   U−B=0.42   V=2985
.SBS3$/   T=3   N00

Only the nuclear region is bright enough to
study here, and it, like NGC 661 above, is
dominated by the light of old stars. Note the
strong reddening in the dust lane. The small
green spot near the nucleus is a plate defect.

NGC 0685

1413 6 81 11 02 04.6   LC 1.0m
B=12.0*   B−V=0.45   U−B=−.20   V=1242
.SXR5..   SBc(r)II   T=5   N00

A weak yellow extension is seen along the
short bar. Star formation activity is distributed
rather uniformly throughout the spiral pattern.

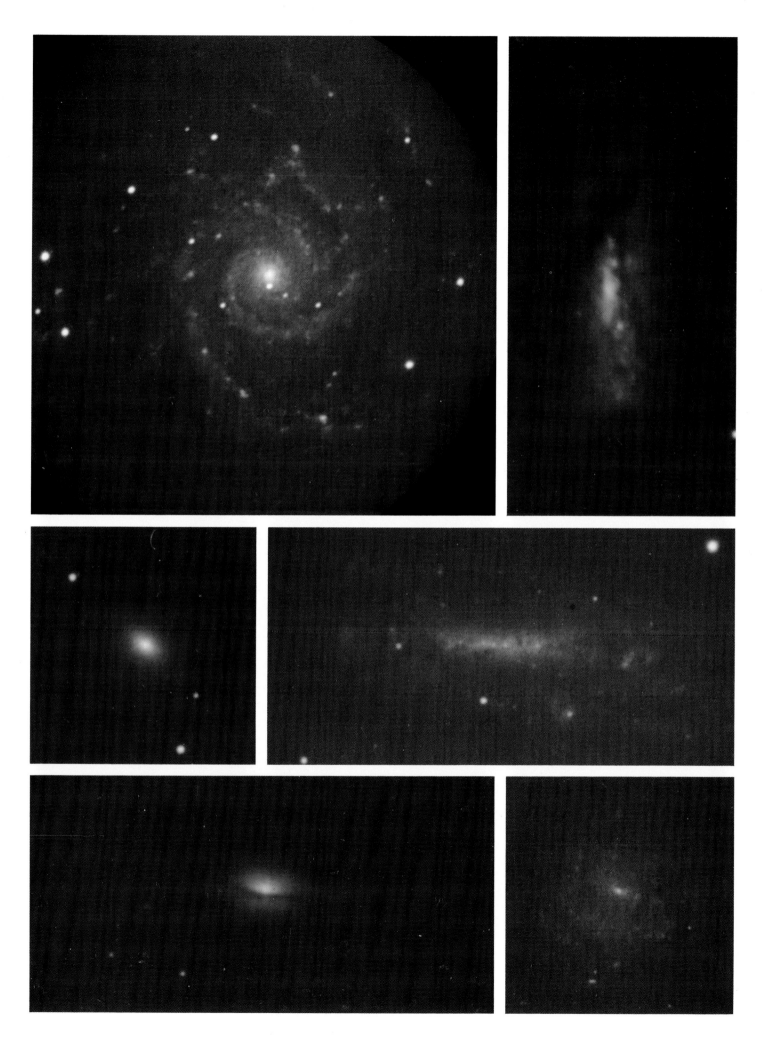

0691
0694
0696
0772
0784
0891
0895

NGC 0691

940 3 79 11 25 06.0   McD 2.7m
B=12.4   B−V=0.79   U−B=0.23   V=2811
.SAT4..   T=4   N17

The blue spiral arms are faint, barely above the threshold for detection. A single bright blue stellar object is seen at 02,3. If it is a blue knot in NGC 691 then it is a unique feature in that galaxy. A faint blue diffuse object is also seen at 09,7. This object appears to be a dwarf elliptical companion with a metal-poor old stellar population.
Corwin has obtaincd photoelectric photometry of the two bright stars: B*1 (left) V=10.97, B−V=0.49, U−B=−0.01; B*2 (right) V=10.85, B−V=0.51, U−B=0.05.

NGC 0697

941 3 79 11 25 06.2   McD 2.7m
B=12.7   B−V=0.77   U−B=0.04   V=3253
.SXR4*.   T=4   N18

Although the old yellow population produces the dominant color in the central disk, the patchy blue structure of the younger spiral features can be seen penetrating deeply into the nuclear region. Compare the overall system color and the appearance of the blue features embedded in the old yellow population with those of NGC 660, with regard to the comments offered on that galaxy.
Two faint diffuse images near the bottom edge right (13,6) are phosphor decay images from the two bright stars in the NGC 691 field above, which was the previous plate set. The decay interval is 12 minutes. Other occurrences of decay images are noted.

NGC 0784

942 3 79 11 25 06.5   McD 2.7m
B=12.2   B−V=0.50   V=362
.SB.7*/   T=7   N27

A faint hint of yellow reveals the presence of older stars distributed throughout the central area of what might otherwise be a candidate for an intrinsically young galaxy. The system does not appear to have a nucleus. Nevertheless, there is a significant smoothly distributed population component. It would be interesting to know if the blue color of that population is influenced more strongly by low metallicity in this case than by youth. If so this could be one of the most slowly evolving galaxies in the atlas.

NGC 0891

458 6 78 01 08 04.1   McD 0.7m
B=10.9   B−V=0.92   U−B=0.27   V=706
.SAS3$/ Sb on edge   T=3   N28

Note the extreme reddening immediately adjacent to the dense dust lane. Despite this dust many blue knots are visible in the outer disk.

NGC 0895

1458 6 81 12 18 04.7   McD 0.7m
B=12.3   B−V=0.54   U−B=−.05   V=2319
.SAS6..   Sc(r)I   T=6 L=3 N00

This system appears to possess a short yellow bar.

NGC 0694

429 6 77 11 06 07.7   McD 2.1m
B=13.8   B−V=0.55   U−B=−.20   V=3017
.L...$P   T=2   N00

There is so much star formation activity in the central region of this galaxy that it is barely possible to detect the old stellar population there. Of the three brightest knots in the nuclear region, the upper one is perceptibly yellow and can be identified with the nucleus. Outside the region of intense activity is a relatively faint, smooth appearing disk apparently composed mostly of stars of old to intermediate age. This disk appears to have better symmetry with the nucleus than do the active regions. Compare with NGC 278 and NGC 3032, and see discussion accompanying NGC 4826.

NGC 0772

475 6 78 01 09 02.7   McD 0.7m
B=11.1   B−V=0.77   U−B=0.05   V=2562
.SAS3..   Sb(rs)I   T=3 L=1 N06

Blue spiral features extend well into the central lens. The singularly bright and narrow spiral arm is the bluest region in the galaxy. Note that even in such a linearly arrayed feature the blue knots still appear as very compact regions with no evident extension along the long arm axis. Clearly the rate of propagation of the star formation signal along the arm axis (along the density wave induced compression zone) cannot proceed faster than the lifetime of a blue knot per blue knot diameter. Hence the brightest stars, those responsible for the blue knot luminosity, must die at least as fast at one edge of a blue knot as the star formation signal is being interpreted at the other edge to start new star formation. Or, in another scenario, the density wave induced compression zone must somehow break up into 'droplets' of a few hundred to a few thousand solar masses which then contract by self-gravitation to form the compact arena in which the star formation occurs. This latter model is at least one step more complex than the first, and moreover suffers from the disadvantage that, at least some of the time, optimum star formation conditions might arise from the shock zones before there has been time for fragmentation to occur, leading to the formation of radically elongated blue knots. Since such features are essentially non-existent in the atlas sample of galaxies the balance of evidence is shifted in favor of the interstellar medium 'digestion' model for blue knot formation, and hence for formation of at least the most massive stars, and perhaps all stars, as evidence derived from 'fossil' spiral arm features will show.

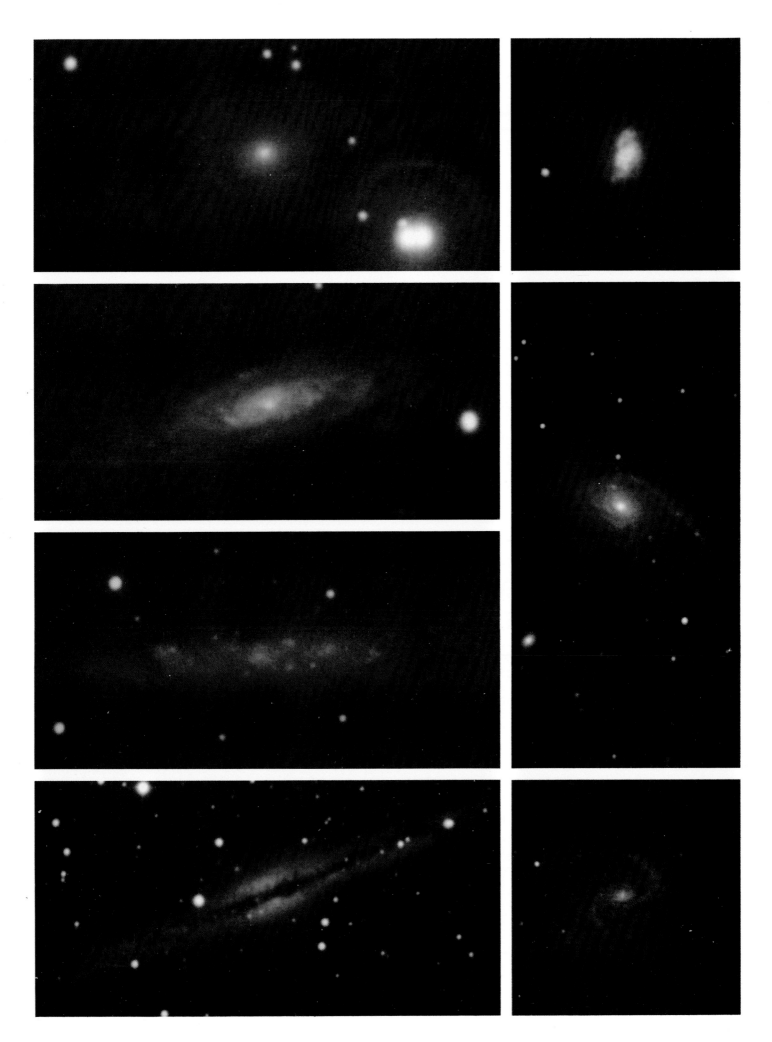

0908
0936
0972
0986
1003
1023

NGC 0908

1324	6	81 10 29 06.2		LC 1.0m
B=10.8	B−V=0.67	U−B=−.05		V=1470
.SAS5.. Sc(s)I−II			T=5 L=1	N35

Outside the immediate nuclear region there is a gradual radial variation in relative strength of intermediate versus young stellar populations. The brightest blue knots are typically compact.

NGC 0936

948	3	79 11 25 07.5		McD 2.7m
B=11.1	B−V=0.96	U−B=0.55		V=1350
.LBT+.. SB02/3/SBa			T=1	N02

Compare the structure of this apparently fully evolved bar and ring structure with the bar and ring system of NGC 151. This comparison may provide information concerning the relative long term stability of three basic morphological structures: bar, ring, and spiral arm. All three are present in a still evolving stellar population (NGC 151), but only the first two are found in the system with the most fully evolved stellar population (NGC 936). In the case of NGC 936 the end regions of the bar are enhanced, and are coincident with the ring structure. This leads to an interpretation of these bar-end enhancements as being libration velocity nodes in the orbits of the stars which comprise the ring. Note the comparison of NGC 936 with NGC 1079 and NGC 1326, for example, and the contrast with NGC 1291.

NGC 1003

943	3	79 11 25 06.7		McD 2.7m
B=12.1	B−V=0.56	U−B=−.12		V=790
.SAS6..			T=6	N34

Older stars are detected in the inner disk. Overall, the galaxy is dominated by young and intermediate age populations. Several galaxies are visible in the background. One of these (at 14,8) appears to be very similar to NGC 1003. Corwin's photometry of the bright star is: V=9.88, B−V=1.01, U−B=0.71.

NGC 0972

1002	3	79 11 26 05.5		McD 2.7m
B=12.1	B−V=0.85	U−B=0.20		V=1670
.I.0... Sbp			T=0 L=3*	N02

Note the diffuse arms of intermediate age, and the young regions of star formation extending well into the nuclear region. Compare with NGC 660 and NGC 697.
Corwin's photometry of the green star: V=9.93, B−V=0.20, U−B=0.15.

NGC 0986

1399	6	81 11 01 05.6		LC 1.0m
B=11.6	B−V=0.71	U−B=0.05		V=1958
.SB.9$.	SBb(rs)II−III		T=9	N09

Note the remarkable similarity with NGC 613. The small spiral structure in the nuclear region exhibits some star formation, and does not appear to be simply the result of dust obscuration although that is a major contributor to the appearence of this structure. The dust lanes along the bar terminate in regions of intensified star formation. The spiral arms bifurcate at these points. In the upper arm a blue knot is located apparently precisely at this point of bifurcation. In this case the inner arm segments form a nearly continuous ring structure. The population of the bright ring segment on the upper right (04,1−08,1) is noticeably yellow. Overall, this galaxy appears to represent a slightly more evolved version of NGC 613.

NGC 1023

459	6	78 01 08 04.7		McD 0.7m
B=10.2	B−V=1.00	U−B=0.55		V=776
.LBT−.. SB01(s)			T=3	N09

This nearly edge-on system is actually very similar to NGC 936. The two regions of enhanced brightness which essentially comprise the bar are seen here at 03,1 and 21,1.
A second galaxy is faintly visible at 35,4. On the original print it is noticeably blue, yet it appears smooth and amorphous with no sign of a yellow nuclear region. From these we conclude that the system is most likely a low surface brightness dwarf elliptical composed mostly of old metal-poor stars. A similar galaxy is seen in the NGC 691 field. See also the companion to NGC 4631.
The field contains several blue stellar objects.

NGC 1042

426	6	77	11	06	07.3		McD 2.1m

B=11.5   B−V=0.55   U−B=−.15   V=1360
.SXT6..   Sc(rs)I–II   T=6   N25

The nuclear region consists of a relatively bright nucleus with a small and much fainter lens. The arm segment at 00,2 appears to be of intermediate age, although it is of moderately high surface brightness. Blue knots are widespread outside of the central region.

NGC 1058

944	3	79	11	25	06.8		McD 2.7m

B=12.1   B−V=0.61   V=674
.SAT5..   Sc(s)II–III   T=5 L=6* N04

The inner disk appears to be a mixture of all age groups.
The lower part of the field contains what appears to be a group of background galaxies of various colors. One object at 20,5 is extremely blue.

NGC 1055

477 6 78 01 09 03.7   McD 0.7m
B=11.4   B−V=0.85   U−B=0.20   V=1077
.SB.3*/   Sbc(s)II   T=3 L=4 N15

Note the nuclear bulge above and below the disk. A number of blue knots are visible in the disk despite the massive amounts of dust. Reddening due to dust is particularly evident in the nuclear region.

NGC 1073

309 6 77 10 17 07.0   McD 2.1m
B=11.5   B−V=0.53   U−B=−.05   V=1245
.SBT5..   SBc(rs)II   T=5 L=3 N01

The bar contains a small sharply defined nucleus and a considerable amount of star formation activity. In contrast, the broad arm-ring feature at the upper left is lacking in star formation regions. Dominated by intermediate age to old stars, this feature seems to be relatively stable. Outside the ring structure the pattern becomes patchy and almost chaotic with numerous regions of active star formation.

NGC 1068   M 71

267b 6 77 10 15 10.5   McD 0.9m
B=9.5   B−V=0.70   U−B=0.08   V=1134
RSAT3..   Sb(rs)II   T=3   N00

The entire central region is too bright to reveal color information in this 'standard' exposure. There appears to be a relatively smoothly distributed yellow population visible throughout the disk. This older population dominates the outer part of the disk. Star formation is seen occurring in the inner part of the disk extending inward to the point where brightness prevents interpretation.
See NGC 694.

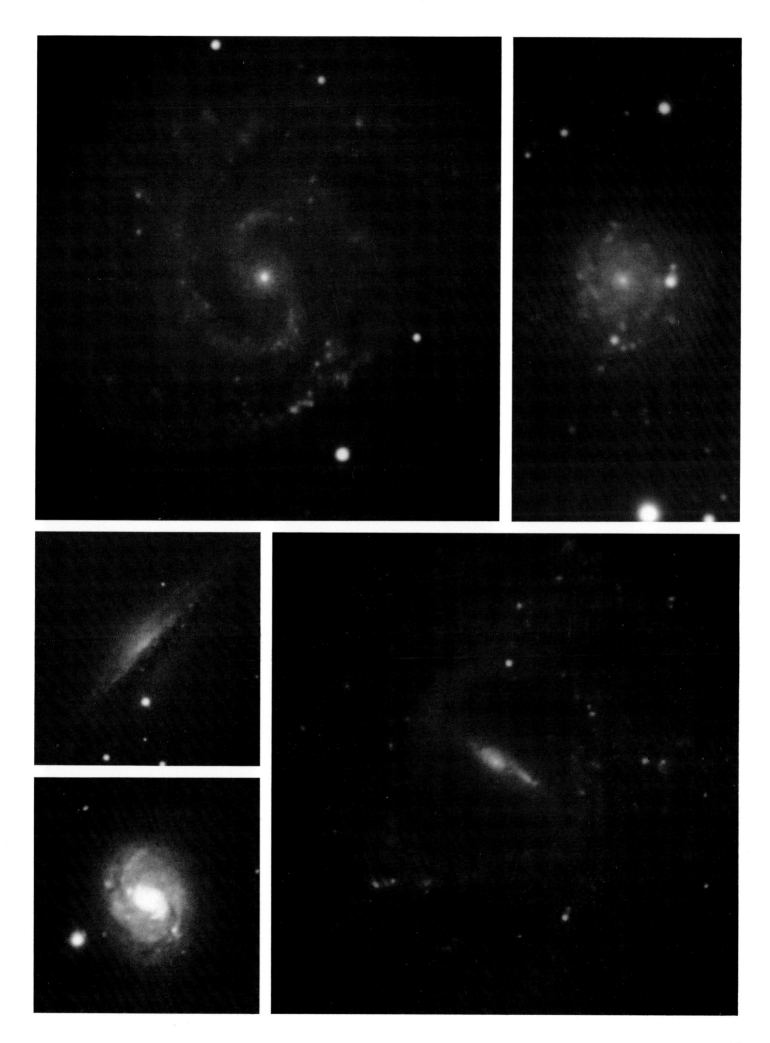

NGC 1079

1375　6　81　10　31　04.8　　　　　　LC 1.0m
B=12.3　　B−V=0.93　　U−B=0.40　　V=2165
RSXT0P.　Sa(s)　　　　　T=0　　　　N00

This system is similar to NGC 936 and NGC 1023, with the bar-end regions extended into short arcs, apparently by libration motion.

NGC 1087

1459　6　81　12　18　05.2　　　　　McD 0.7m
B=11.5　　B−V=0.56　　U−B=−.09　　V=1844
.SXT5..　Sc(s)III.3　　　　T=5　L=5* N25

A short bar contains some old stars, but is dominated more by discrete regions of star formation. Blue knots are scattered over the disk without apparent spiral order. This system is probably a good example of stochastic star formation processes in operation.

NGC 1084

432　6　77　11　06　08.1　　　　　McD 2.1m
B=11.2　　B−V=0.62　　U−B=−.05　　V=1406
.SAS5..　Sc(s)II.2　　　　T=5　L=2 N25

Except for its unusually high surface brightness this galaxy is similar to NGC 1087. The system contains old stars in the nuclear region, but the young and intermediate age populations dominate. The boundary is relatively amorphous, and seems to be of intermediate age. Compare also with NGC 278.

NGC 1087

369　6　77　11　05　08.5　　　　　McD 2.1m
B=11.5　　B−V=0.56　　U−B=−.09　　V=1844
.SXT5..　Sc(s)III.3　　　　T=5　L=5* N25

This is the first of a number of examples in the atlas where the same galaxy is illustrated as observed with different telescopes. The main purpose of these dual illustrations is to provide a further basis for judging the repeatability of the colors , not only from one telescope to another, but from one night to another, for different zenith distances and different air masses, different image tubes and any other parameters that could enter in to produce differences in the final results. You will find that the agreement is on the whole reasonably good, with occasional obvious differences which you should consider in your own interpretation of the information conveyed in these color images. In this case the only noticeable difference is in the bar, which appears somewhat redder in the smaller scale image. This is partially due to the loss of resolution of the blue knots in the bar, but not entirely; and here is where you must assume some responsibility and caution in interpretation. On the whole, I believe you will find these comparisons will give you a properly qualified confidence in the atlas overall.

NGC 1097

1442　6　81　11　04　06.2　　　　　LC 1.0m
B=13.7　　B−V=0.85　　U−B=0.43　　V=1205
.SBS3..　RSBc(rs)II−III　　　　　　N06

The nuclear region exhibits a distinct spiral pattern in which if any star formation is present it is swamped by the apparent overwhelming dominance of an old population in the inner lens. Star formation is seen to be occurring in and near the dust lanes in the bar. A bright region of intermediate to old age is seen at the terminus of the dust lane at 09,3. This may indicate the initiation of co-rotation accumulation leading eventually to bar-end features such as in NGC 1079 (above). The ring structure is relatively blue, comparable to the spiral arms. Note the spatial hiatus between the ring structure at 29,4 and the beginning of the spiral arm. This most likely is not due simply to the presence of some intervening dust lane, but instead represents an absence of star forming materials or a zone of 'low pressure' sufficient to inhibit star formation.

The green spot at 32,1 is a plate defect.

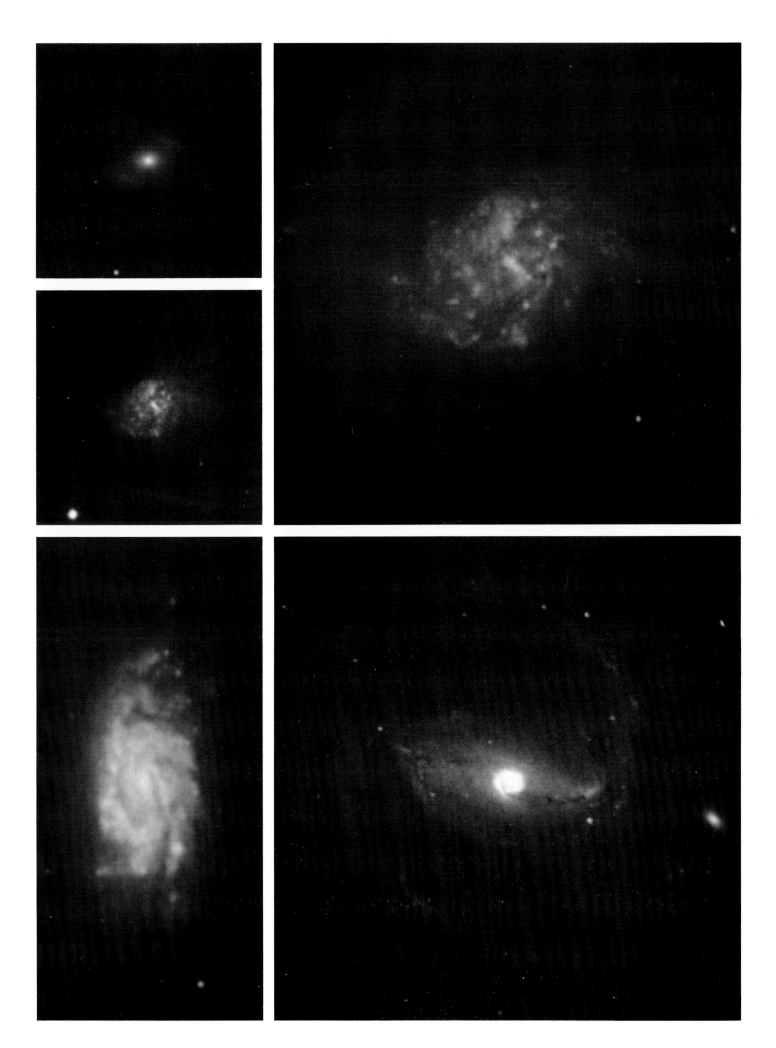

1156
1186
1187
1201
1232
1241
1242
1255

NGC 1156

378 6 77 11 05 09.9   McD 2.1m
B=12.2   B−V=0.56   U−B=−.15   V=485
.IBS9..   SmIV   T=0 L=7* N09

This system does not evidence any yellowing towards the center, hence it may actually be lacking in an old population altogether. It is also possible that an old population is present, but that it is very metal-poor. In any case there is no evidence for the existence of a central concentration of stars as is common to the vast majority of galaxies. Thus this system may be intrinsically young. Note two 'violet knots' which at this resolution are probably best understood as O-associations with unusually large quantities of oxygen in high ionization states, i.e. blue knots containing extremely massive stars.

NGC 1187

1345 6 81 10 30 04.2   LC 1.0m
B=11.1   B−V=0.55   U−B=−.10   V=1334
.SBR5..   SBbc(s)I−II   T=5 L=1 N18

The bar of this system appears to be established, but it is not composed of old stars. We may see in this galaxy a system at a relatively early stage on a definitive evolutionary path culminating in an NGC 936 type system. See also NGC 1241 on this page.

NGC 1201

1414 6 81 11 02 04.8   LC 1.0m
B=11.5   B−V=0.94   U−B=0.50   V=1630
.LAR0*.   S0₁(6)   T=−2   N13

Two faint diffuse patches are seen symmetrically located at 12,1 and 30,1.

NGC 1241

1003 3 79 11 26 06.3   McD 2.7m
B=12.7*   V=4042
.SBT3..   SBbc(s)I.2   T=3   N29

The bar of NGC 1241 (08,1) is better defined and contains a greater percentage of old stars than the bar in NGC 1187 (above), and the ring is brighter with respect to the outer spiral features, indicating that NGC 1241 is slightly further evolved than NGC 1187, although both appear to be on similar evolutionary paths.

The smaller galaxy, NGC 1242, also evidences a bar structure in yellow light, although the older population does not yet dominate that structure completely.

Corwin's photometry of the bright star: V=9.30, B−V=0.99, U−B=0.67. A galaxy with similar photoelectric colors typically appears yellow (see for example NGC 1201

NGC 1186

946 3 79 11 25 07.1   McD 2.7m
.SBR4*.   T=4 N17

There is at present relatively little data on this galaxy. It has a well established bar of old population. The ring and spiral structures contain significant star formation activity. Two other galaxies seen in the field appear blue. The galaxy at 11,7 may be a companion similar to the companion to NGC 4631. If the galaxy at 08,5 is a background object, its red-shift-corrected color must be extremely blue. The faint red flare at the lower left is an artifact.

NGC 1232

478 6 78 01 09 04.0   McD 0.7m
B=10.5   B−V=0.63   U−B=−.05   V=1644
.SXT5..   Sc(rs)I   T=5 L=1 N00

This 'unusually normal' system has some of everything, all in the right places. The color gradient is from yellow to blue. The spiral arms are reasonably distinct, not too open, and not too tightly wound; and they are composed of representative samples of blue knots and intermediate age stars. Are such systems rare because they do not last long, or is it because the initial conditions for galaxies are widely variant and lack the 'peak' of a normal distribution in the probability of their (particular initial conditions) occurrence? The complex at 25,6 appears to have some degree of internal structural independence, and if the bright feature is a bar structure as it appears to be, despite its blue color, then it is indeed an independent dynamical system. A third galaxy, probably a background object, but still relatively blue in color, is visible at 00,2.

immediately above). Both stars and galaxies follow color sequences that are shifted systematically by about 0.2 magnitude in apparent color. This is not a bias in printing as can be ascertained by comparing the colors of similar regions in different galaxies, such as the NGC 1241 with NGC 1201. The old popularion seen in the nuclear regions of galaxies appears rather similar for all the galaxies on this page as it does throughout the atlas in general; nowhere does this old population assume the color of this star despite the fact that their photoelectric colors are virtually identical. This effect must be due to differences in the instrumental response between the image tube UBV system and the photoelectric UBV system, together with differences in the energy distribution for stars and galaxies within each of the passbands.

NGC 1255

1326 6 81 10 29 06.7   LC 1.0m
B=11.6   B−V=0.52   U−B=−.10   V=1738
.SXT4..   Sc(s)II   T=4 L=3 N18

Despite its low surface brightness, and thus, if we may presume, low mass, the dynamical structure of NGC 1255 does not differ greatly from that of the somewhat brighter NGC 1187. Note the singularly bright unresolved blue knot at 32,2.

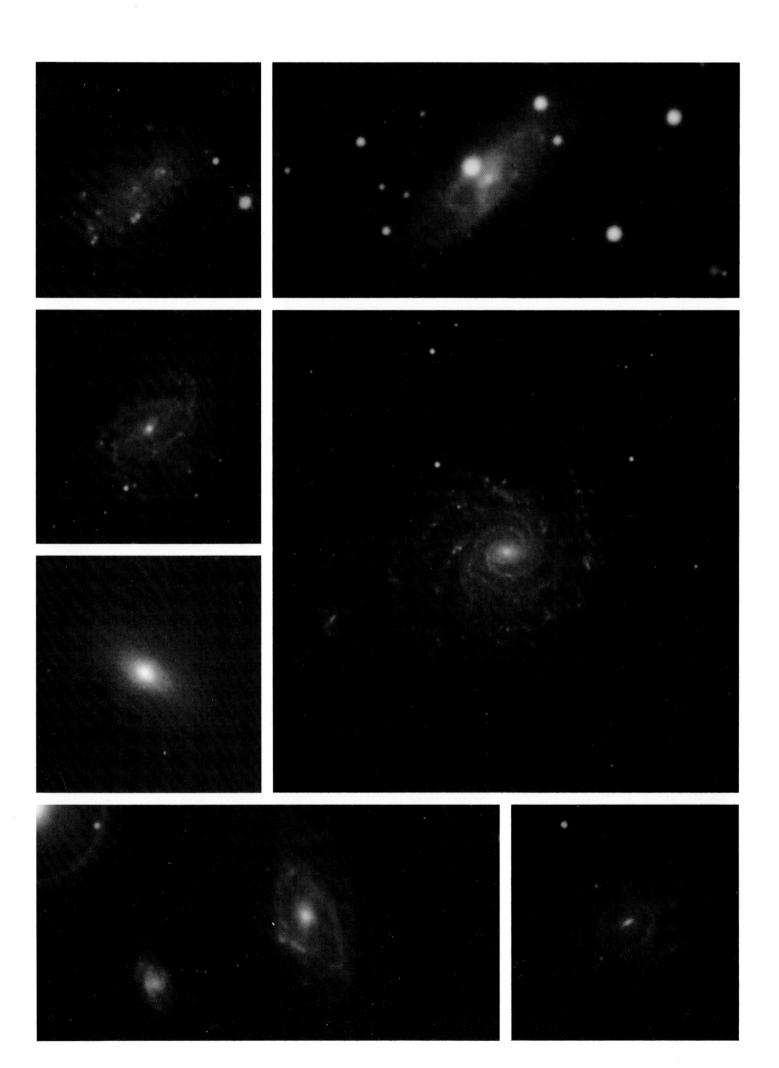

1275
1288
1291
1300
1302
1309
1310
1313

NGC 1275

947 3 79 11 25 07.3			McD	2.7m
B=12.3	B−V=0.76	U−B=0.10		V=5361
.P.....	E pec		T=2	N00

The faint spiral structure of a very late type galaxy can be traced in the original print as two blue arms. The nucleus of the spiral galaxy is the small yellow object superimposed on the elliptical galaxy at 30,1. In the atlas reproduction only the brighter of these blue regions in the spiral galaxy are seen. The discernment of the nucleus of the spiral galaxy makes it clear that the NGC 1275 object actually comprises two galaxies, apparently in collision (or at least superimposed in the line of sight) rather than a single exceedingly peculiar elliptical galaxy.

NGC 1300

457 6 78 01 08 03.6			McD	0.7m
B=11.1	B−V=0.68	U−B=0.13		V=1422
.SBT4..	SBb(s)I.2		T=4 L=1	N00

The comparatively low surface brightness of NGC 1300 is evident in this photograph. Star formation regions occur near the ends of the bar and in the arms. Leading edge dust lanes are just visible in the bar. A faint libration node enhancement is seen at 09,2 as a diffuse yellow patch at the end of the bar.

NGC 1309

1004 3 79 11 26 06.4			McD	2.7m
B=12.0	B−V=0.45	U−B=−.18		V=2195
.SAS4*.	Sb(rs)II		T=4 L=4	N00

The predominant morphological feature of this galaxy is its plume structure. The primary spiral arm structure is nearly lost in the wide-spread system of plumes. From an evolutionary perspective we note that the galaxy is dominated by a present burst of star-formation activity which, on the basis of the general lack of any significant yellow disk population, may be unprecedented in its evolutionary history.
A small blue galaxy is visible at 31,3.

NGC 1310

1415 6 81 11 02 05.2			LC	1.0m
				V=1574
.SAS5*.			T=5	N00

The nuclear region appears slightly decentered with respect to the surrounding fainter yellow disk. Several unresolved blue knots are visible near the outer boundary of the disk as seen in this photograph.

NGC 1288

1346 6 81 10 30 04.5			LC	1.0m
B=12.8	B−V=0.69	U−B=0.05		V=4371
.SAS4..	Sab(r)I−II		T=4 L=1	N00

This galaxy appears similar to NGC 1232. It is interesting to note here that the red shift velocities agree very closely with the apparent angular sizes of these galaxies if both are the same intrinsic size.

NGC 1313

1328 6 81 10 29 07.3			LC	1.0m
B=9.3	B−V=0.42	U−B=−.23	V=241	
.SBS7..	SBc(s)III−IV		T=7	N04

The nucleus appears distinctly green rather than yellow. The disk population, identified by its relative smoothness, also appears green. The remainder of the light from this system originates almost entirely from discrete blue knots, one group of which (03,2) is brighter than the nuclear region itself. Does this system contain even a single old star? We know that the older stars tend to be more smoothly distributed than the young ones. If young stars are formed in regions constrained by the particular requirements for star formation (gas and dust present in sufficient density) are these regions smoothly distributed? Certainly not in the case of the blue knots which are the birth sites for at least the massive stars, those which give the blue knots their color. Furthermore we know that star clusters contain stars of all masses, not just the most massive stars; therefore is it necessary to postulate star formation outside of the blue knots at all? We can examine these factors with regard to the green disk on NGC 1313. The possibilities are: (i) The smooth disk contains only stars of intermediate age, whose intrinsic color is green. There are no old stars in the system at all. (ii) The disk contains a smooth distribution of old yellow stars and a sufficiently smooth distribution of young stars

NGC 1291

1327 6 81 10 29 07.1			LC	1.0m
B=9.4	B−V=0.93	U−B=0.44		V=674
RSBS0..			T=0	N00

This system lacks libration node enhancements at the bar ends. No dust features are evident. The dark spots at 1,5 and 1,12 are due to defects present in the detector system during the Las Campanas observing run of October–November 1981.

NGC 1302

1400 6 81 11 01 05.8			LC	1.0m
B=11.1	B−V=0.85	U−B=0.33		V=1626
RSBS0..	Sa		T=0	N00

This galaxy is dominated by the yellow (old) stellar population. It is difficult to determine here if the patchiness is intrinsic to the galaxy, or is simply due to detector noise. Compare with NGC 7743.

(blue stars) not in blue knots to give the appearence of a uniform green color to the smooth disk. (iii) The disk contains old metal-poor stars whose colors in the UBV system are shifted appreciably towards the blue from the normal yellow of a typical old population. (iv) The color-balance of the original dye transfer image was off enough for yellow to appear green. You must verify for your self that (iv) is not likely. Compare the apparent colors of galaxies throughout the atlas with the photoelectric data in the headers. Compare galaxies of similar morphology for differences in color. In particular compare the colors of the smooth population in the irregular galaxies. (See NGC 2188, 2552, 3395-96, 3644, 4214, 4449, and 4490 for example.) There seems little doubt that these smooth populations are indeed green. The metal-poor hypothesis offers a likely mechanism. Indeed, several small elliptical galaxies are seen to be blue (see for example the companion to NGC 4631), and it is assumed that metal deficiency is the cause, although it may not be. If the disk regions of irregular galaxies tend to be metal deficient, then this tells us that the rate of infusion of heavy elements into the interstellar medium of irregular galaxies is (or was) abnormally low in comparison to the rate of *(text continues after NGC 1365)*

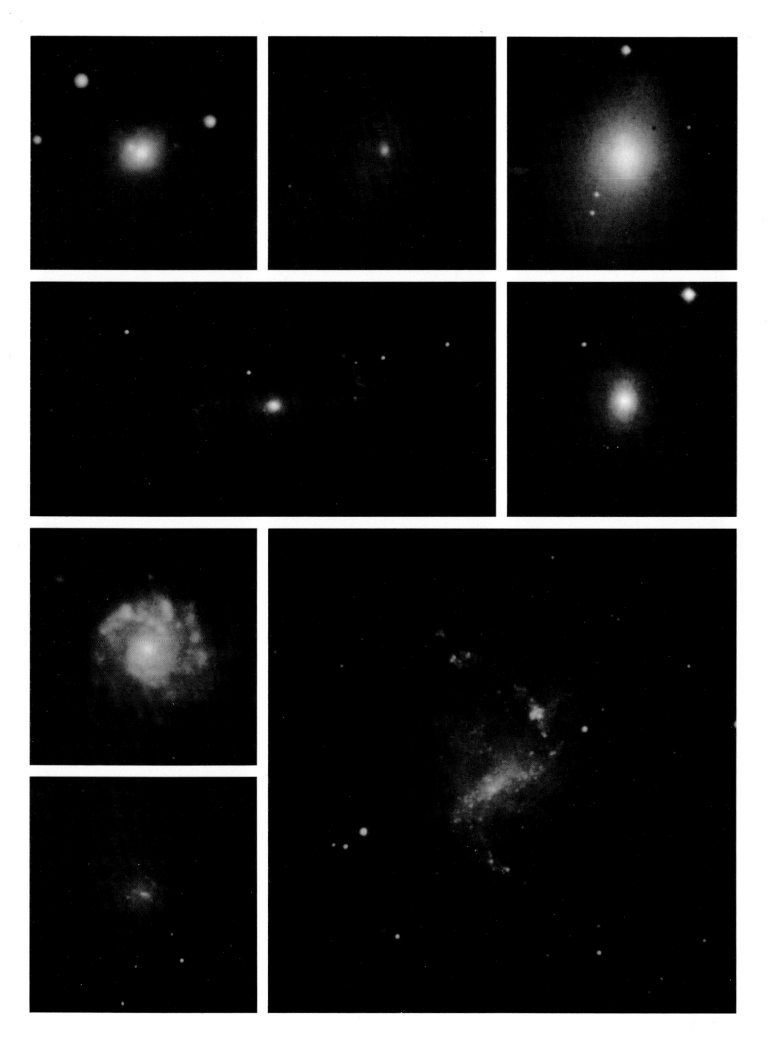

NGC 1316 Fornax A

1401	6	81 11 01 06.1			LC 1.0m
B=9.6	B−V=0.90		U−B=0.49		V=1632
PLX50P.	Sap merger?		T=−2		M32

A chain of dust clouds crosses the minor axis, suggesting a prolate system with an equatorial dust ring. Compare with NGC 5266. This image contains the same two defects present in the NGC 1291 field. Here they are at 06,1 and 01,2.

NGC 1317

1350	6	81 10 30 05.7			LC 1.0m
B=11.9	B−V=0.93		U−B=0.30		V=1918
PSXT1..	Sa		T=1		N00

A ring of blue knots surrounds the bright yellow lens. A faint yellow bar extends beyond the blue ring.

NGC 1325

1380	6	81 10 31 05.8			LC 1.0m
B=12.3	B−V=0.77		U−B=0.07		V=1543
.SAS4..	Sb		T=4		N00

A small yellow nucleus and a single unresolved bright blue knot punctuate this galaxy. The faint disk appears to be composed mostly of stars of intermediate age.

NGC 1326

1402	6	81 11 01 06.4			LC 1.0m
B=11.3	B−V=0.82		U−B=0.27		V=1167
RLBR+..	RSBa		T=1		N32

Note that the inner bar axis is rotated with respect to the axis defined by the outer bar enhancements. Compare with NGC 936. This system is at a very late stage of evolution.

NGC 1326b

1403	6	81 11 01 06.7			LC 1.0m
					N32

This system is apparently composed almost entirely of very young stars. This may be an intrinsically young galaxy. Compare with NGC 3109.

NGC 1337

1433	6	81 11 03 06.3			LC 1.0m
B=12.2	B−V=0.56		U−B=−.05		V=1189
.SAS6..	Sc(s)I−II		T=6	L=7*	N05

The small yellow lens is interpenetrated with blue knots. The brightest blue knots are located in the inner region of the disk. The outer disk region appears faint, patchy, and intermediate in color.

NGC 1353

1416	6	81 11 02 05.4			LC 1.0m
B=12.2	B−V=0.90		U−B=0.28		V=1605
.SBT3*.	Sbc(r)II		T=3	L=4	N00

The old yellow population dominates, but a number of star formation regions are visible in the ring-like arms.

NGC 1357

1006	3	79 11 26 06.8			McD 2.7m
B=12.6	B−V=0.93		U−B=0.26		V=1929
.SAS2..	Sa(s)		T=2		N00

The spiral arms appear to begin well outside of the lens, and tangent to it. A bright star formation region is seen near the end of one arm at 35,2.

NGC 1358

950	3	79 11 25 07.9			McD 2.7m
B=13.0	B−V=0.94		U−B=0.45		V=4003
.SXR0..	SBa(s)I		T=0	L=3*	N00

The bar is somewhat small with respect to the scale of the spiral arms. Compare this spiral structure with that of NGC 210. The bar structures are quite different, however. Neither exhibits ring structure of the NGC 151 type.

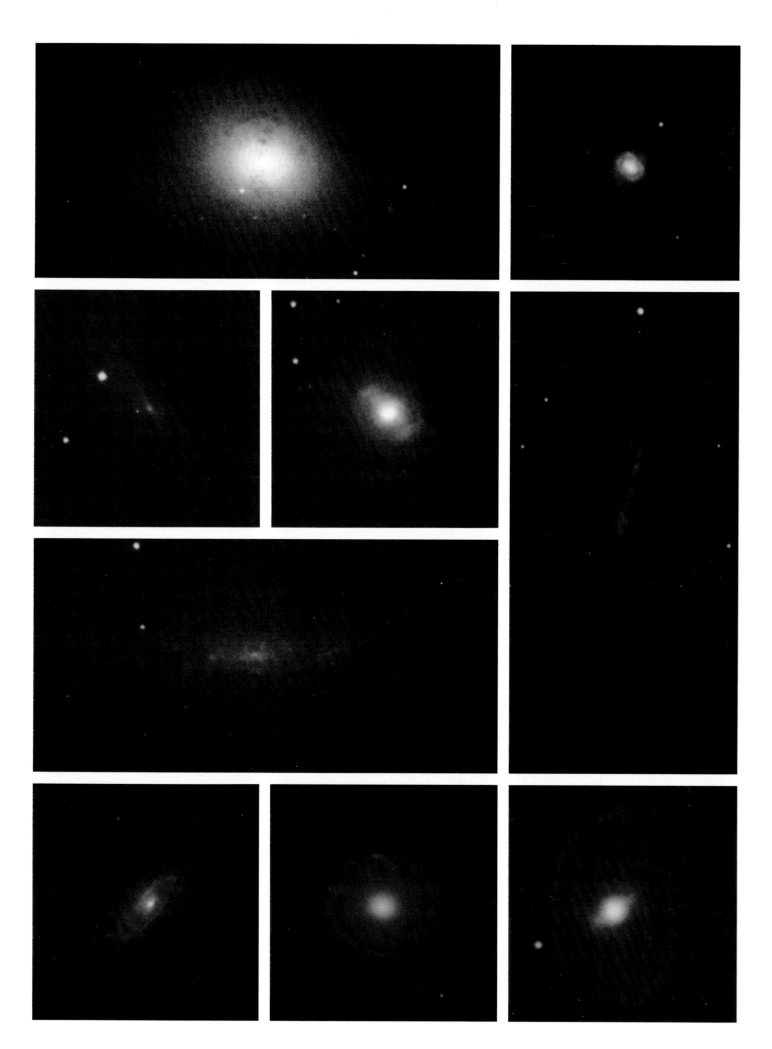

1404 6 81 11 01 07.0   LC 1.0m
B=10.1   B−V=0.62   U−B=0.05   V=1502
.SBS3..   SBb(s)I   T=3 L=2 N33

The nuclear region is dominated by a spiral pattern of dust and bright knots. Although the dominant color there is yellow, blue knots are visible along the leading edge of the dust lane at 30,1, indicating that star formation may be going on well into the nuclear region. The morphology of the nuclear region appears comparable to that of NGC 613, NGC 986, and NGC 1530. The first two of these galaxies have well developed ring structures, and NGC 1530 exhibits traces of a weak, partially developed ring. NGC 1365, however, has no ring, but rather a system of plumes, some of which contain blue knots that may represent the initial stages of development of a ring. The dynamical complexity of a galaxy such as this is two-fold: (i) The dynamics of the visible system and (ii) The dynamics of the density wave system. It is tempting to look at the visible system and contemplate its motion as if that were the essential dynamical nature of the galaxy, whereas to contemplate the tremulous nature of the denstiy wave structure, for which even the instantaneous condition (mass distribution) must be largely inferred, is a difficult problem. It is possible, however, to gain insight into the dynamical process from a careful consideration of the visual appearance of the galaxy provided several key principles are kept in mind. First, the mass to light ratio

for the luminous blue stars is only about .01 of the mass to light ratio for the sun, whereas the mass to light ratio for the yellow population may be 10 or more times the mass to light ratio for the sun. The mass to light ratio for the green appearing regions is intermediate to these two values. Thus, while the yellow regions may be directly associated with the mass distribution picture, the mass associated with the blue knots is totally negligible, and the smooth green regions in the spiral arms contribute only weakly to the systemic mass distribution. The second consideration is the fact that the blue knots are extremely short-lived phenomena (on the galactic time scale), while the yellow regions are comprised of older stars, and the green regions are comprised of stars generally of intermediate age. Third, the density wave does not travel at the same speed as the stars which comprise the spiral arms. Instead, the stars move through the wave, and stars born in the wave's compression zone move out of the wave along with the stars moving through it. By combining these factors, then, you arrive at the realization that the spiral structure defined by the blue knots represents in effect a stop-frame image of the density wave simply because the blue knots do not last long enough to drift appreciably away from their place of origin.

Finally, it is the propagation of the wave structure, together with the nature of the medium in which it is propagated (stars, gas and dust in varying amounts) which determines the dynamical ephemeris of a galaxy; and this combined with the effects of stellar evolution throughout the system gives rise to the pattern of systemic evolution as a whole.

Even with enormous expenditures of human effort and computer resources, present theoretical understanding of density wave dynamics in galaxies remains limited to relatively simple models, while the galaxies themselves remain incredibly complex. As yet, for example, we have no dynamical model whatsoever to account for plume structure, and yet this is the dominant structure (particularly with respect to populations of intermediate age) in many galaxies, as it is in the inner-disk region of NGC 1365. In this arena, then, we are limited to the basic tools of observation, interpretation, inference, and speculation. Based on these considerations it appears reasonable to suggest that this system is very probably evolving towards a ring stage similar perhaps to the present appearance of NGC 5921 or NGC 613.

NGC 1313 contd

star formation. This in turn would imply a low supernova rate and hence an upper limit to the mass of individual stars formed in these galaxies which is systematically less than the mass of supernova progenitors. And yet the blue knots in these galaxies appear qualitatively similar to the blue knots in galaxies containing definite yellow populations in their disks. Hence these photographs themselves do not provide the necessary evidence to establish that (iii) is the explanation for the green disk. With regard to (ii), some faint blue knots are seen superimposed on the disk. To argue that faint blue knots somehow cover the disk uniformly, when they do not appear outside the disk uniformly (in general smooth areas of the same blue color as the blue knots are not found in any galaxies in the Atlas) is difficult to accept. In short, if one component of an observed

smooth green is to be smooth yellow, the other component must be smooth blue, and smooth blue is not found. Perhaps star formation is occurring outside the blue knots. This is not a necessary prerequisite for low mass stars as was pointed out above, but suppose it is nevertheless taking place. To explain the smooth green distribution in this way requires that the smoothly distributed star formation activity (blue) would have to be in essentially precise correspondence to the volume emissivity of the yellow light of the old population. For this to occur in even a single galaxy would demand a causal relationship between the smoothly distributed old population and the phenomenon of star formation. If such a relationship existed it would be exceedingly difficult to account for its absence in the great majority of galaxies which are seen to contain a yellow old

population and not a green one. Thus it is almost certain that the green regions of these galaxies do not contain a 'normal' (yellow) old population of stars. Hence the possibility that these galaxies (ones with green regions but not yellow regions) do not contain any old stars at all. If (i) is to hold, then these systems were formed at a later epoch than other galaxies. Since the green color for stars persists through spectral type G in this color system, this would still permit a wide range in age for galaxies in which the smooth green population represents the oldest population in the system.

In either case it seems most likely that a proper explanation and understanding of these smooth green disks will have to be based on a much more detailed understanding of the formation and evolution of these galaxies than is presently available.

NGC 1371

```
1352  6  81 10 30 06.3              LC 1.0m
B=11.3    B−V=0.92   U−B=0.45        V=1397
.SXT1.. Sa(s)              T=1         N00
```

Only the nuclear region with its old yellow population is bright enough to be recorded here.

NGC 1376

```
951  3  79 11 25 08.1              McD 2.7m
B=12.8*                              V=4128
.SAS6.. Sc(s)II          T=6 L=4* N27
```

A relatively complex multi-armed system. The blue knots appear strongly constrained to the spiral features, an effect more in accord with a wave driven process than a purely stochastic one.

Two faint galaxies, one blue at 06,6 and the other red at 08,5, are visible.

NGC 1380

```
1383  6  81 10 31 06.8              LC 1.0m
B=11.1    B−V=0.94   U−B=0.43        V=1664
.LA....   S03(7)/Sa        T=2         N05
```

The system is dominated by an old yellow population.

A blue stellar object is visible at 10,1.

NGC 1398

```
1417  6  81 11 02 06.0              LC 1.0m
B=10.6    B−V=0.95   U−B=0.40        V=1299
PSBR2.. SBab(r)I           T=2  L=1 N05
```

Dark bands cross the bar forming a quasi bar-lens boundry. The ring contains several faint blue knots and enough young to intermediate age stars to appear relatively blue in overall color. The galaxy's faint spiral structure is just below the limit of visibility in this reproduction.

NGC 1385

```
1406  6  81 11 01 07.6              LC 1.0m
B=11.6    B−V=0.49   U−B=−.20        V=1389
.SBS6.. ScIII            T=6  L=2 N27
```

Note the white nuclear region. Compare with NGC 694. Photoelectric colors of NGC 55 are also similar, but the surface brightness is low enough for the actual color to be evident.

NGC 1415

```
1434  6  81 11 03 06.6              LC 1.0m
B=12.0    B−V=0.88   U−B=0.30        V=1399
RSXS0.. SaSBa late       T=0  L=5 N00
```

A broad flat bar is seen viewed from an oblique and nearly end-on perspective. Note the typical dust lanes and the star formation activity generally near the ends of the bar.

NGC 1417

```
952  3  79 11 25 08.2              McD 2.7m
B=12.7    B−V=0.72   U−B=0.15        V=4057
.SXT3.. Sb(s)I.3           T=3 L=2* N05
```

A spiral pattern is evident in the yellow nuclear region. Star formation activity is particularly strong immediately adjacent to the lens in the inner disk region, and declines noticeably in the outer arms.

NGC 1418

```
1007  3  79 11 26 06.9             McD 2.7m
B=14.4    B−V=0.70   U−B=0.25        V=3679
.SBS3*.                    T=3         N00
```

A foreground star appears superimposed on the galaxy.

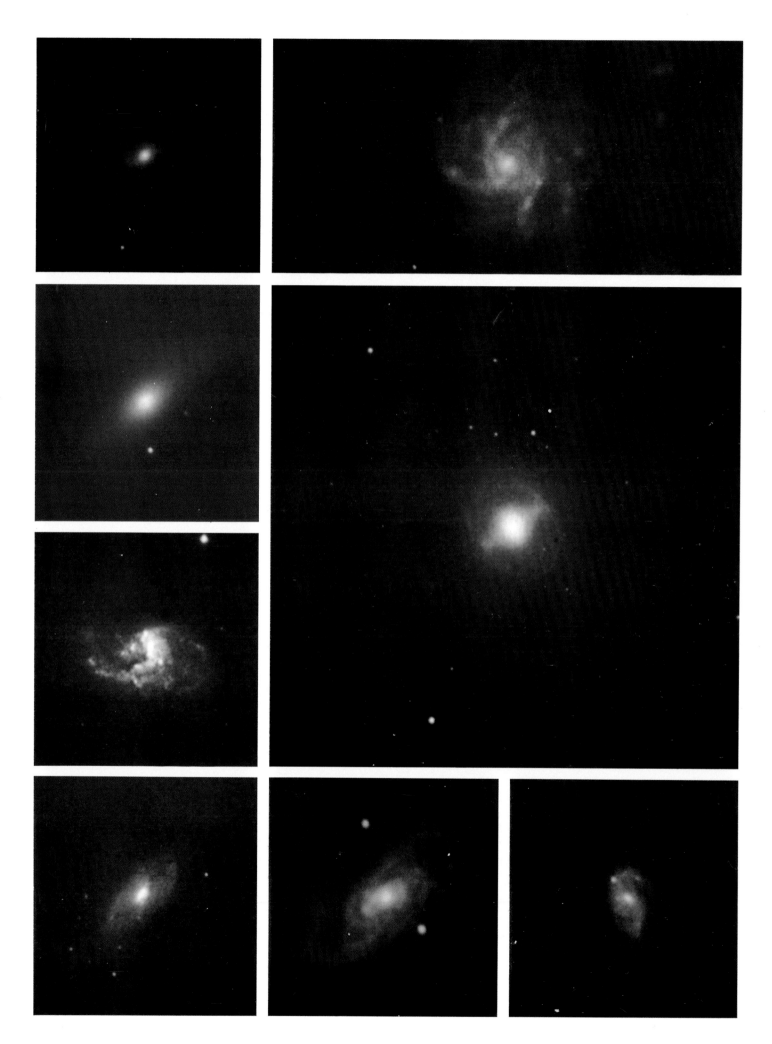

1421
1424
1425
1433
1448
1453
1493
1494

NGC 1421

1005	3	79 11 26 06.6		McD 2.7m
B=11.9	B−V=0.51	U−B=−.02	V=2067	
.SXT4*.	ScIII:		T=4 L=1* N09	

Beyond the small yellow nuclear region the
galaxy is dominated by star formation and
stars of intermediate age, although the inner
disk also has a significant yellow component.
One spiral arm is particularly noteworthy for
the string of nearly contiguous individual blue
knots which extends well into the inner disk.

NGC 1424

1008	3	79 11 26 07.1		McD 2.7m
				V=5595
.SXT3*.			T=3	N31

The large zone of reduced surface brightness
between the lens and the inner spiral structure
suggests the presence of a weak bar not visible
at this resolution. See for example NGC 2545.

NGC 1425

1330	6	81 10 29 07.9		LC 1.0m
B=11.1	B−V=0.70	U−B=0.13	V=1494	
.SAS3..	Sb(r)II		T=3 L=4 N22	

Here we see a relatively smooth transition
from a dominant yellow nuclear region and
lens into a faint multi-armed spiral pattern in
the disk. A number of faint blue knots are
visible in the disk, mostly near the lens-disk
interface.

NGC 1433

353	6	81 10 30 06.6		LC 1.0m
B=10.6	B−V=0.69	U−B=0.23	V=802	
.SBR1..	SBb(s)I−II		T=1	N27

The nuclear region appears to contain a ring of
star formation regions, but lacks the spiral
pattern of dust found in otherwise similar
galaxies. See for example NGC 4314. The bar
appears smooth and yellow, but unusually
faint.
A blue stellar object is seen at 16,7.

NGC 1448

1384	6	81 10 31 07.1		LC 1.0m
B=11.3	B−V=0.70	U−B=0.00	V=1005	
.SA.6*/	Sc(II):		T=6 L=5 N13	

The disk contains sufficient dust to cause
noticeable reddening and extinction to the
nuclear region. If the blue knots are used to
define the disk, and if the nucleus is centrally
located in the disk, then the nucleus itself
would appear to be the very red object
immediately below the bright yellow region.

NGC 1453

1009	3	79 11 26 07.2		McD 2.7m
B=12.6	B−V=1.03	U−B=0.62	V=3861	
.E.2+..	E2		T=−5	N00

A typical yellow old population. Note the
corresponding photoelectric colors above.

NGC 1493

1407	6	81 11 01 07.8		LC 1.0m
B=11.8	B−V=0.53	U−B=0.00	V=835	
.SBR6..	SBc(rs)III		T=6	N00

A yellow nucleus is present. The bar is
comprised of individual blue knots. Scattered
blue knots form a loose ring surrounding the
relatively smooth inner disk.

NGC 1494

1418	6	81 11 02 06.3		LC 1.0m
B=12.2*				V=906
.SAS7..	Scd(s)II		T=7	N27

There is no direct evidence for a nucleus. A
weak old population is present in the central
region. Star formation activity is widespread.

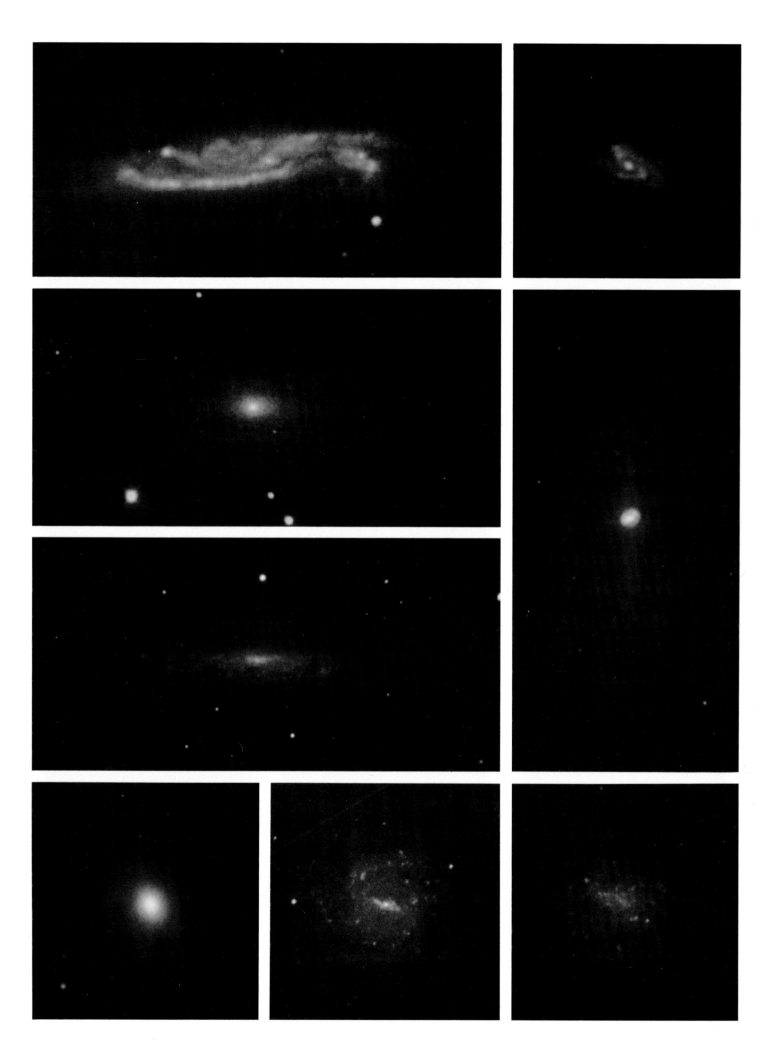

NGC 1507

953 3 79 11 25 08.4   McD 2.7m
B=12.7   B−V=0.51   U−B=−.20   V=850
.SBS9P$ Sc   T=9   N09

The stellar population is mixed, with the old
population predominant in the central region
and with the remainder apparently influenced
mostly by stars of intermediate age. Several
blue knots are evident, with one region of star
formation having unusually high surface
brightness.

NGC 1515

1331 6 81 10 29 08.2   LC 1.0m
B=11.8   B−V=0.85   U−B=0.23   V=924
.SXS$..   Sb(s)II   T=3   N31

Considerable reddening is evident in the lower
third of the bright yellow region. Several blue
knots are visible in the disk.
One very blue stellar object is seen at 31,4.
Several other less pronounced blue stellar
objects are also visible.

NGC 1530

480 6 78 01 09 05.3   McD 0.7m
B=12.1   B−V=0.80   U−B=0.10   V=2701
.SBT3..   T=3   N00

Massive dust lanes range along the bar from
the nuclear spiral to the spiral arms. One dust
lane terminates in a bright blue knot at 08,1.
The yellow bar appears deeply reddened in the
vicinity of the dust lanes. Refer to the
illustration on the right.
A blue stellar object is present at 07,4. The
object at 06,2 is a defect.

NGC 1512

1435 6 81 11 03 06.9   LC 1.0m
B=11.5   B−V=0.84   U−B=0.22   V=558
.LBR+..   SBb(rs)Ip   N23

The nuclear region exhibits the same sort of
spiral pattern visible in NGC 613, NGC 1097
and NGC 1530, although it is less
pronounced, and only one dust lane is clearly
evident. The occurrence of this type of
phenomen within a wide range of galaxy types,
from NGC 1365 to NGC 4314 for example,
suggests an underlying process more
fundamental to the intrinsic physics of galaxies
than to their outward appearance.
NGC 1510 is at the top of the field.

NGC 1530

441 6 77 11 07 06.8   McD 2.1m
B=12.1   B−V=0.80   U−B=0.10   V=2701
.SBT3..   T=3   N00

At this larger scale several additional blue
knots are seen in the dust lane complex along
the bar. Despite the less than perfect seeing the
spiral pattern in the nuclear region is resolved
sufficiently to reveal blue knots within the
nuclear spiral. The direct observation of
groups of young stars existing in regions
overwhelmingly dominated by old stars, not
only in many different galaxies, but in many
different types of galaxies as well, is another
indication that models for the evolution of
galaxies must take into account not simply
enrichment of the interstellar medium but its
complete replenishment, at least in certain
restricted regions.
Comparing this photograph with the one on
the left, taken two months later with a
different telescope, it is evident that, in this
case at least, the color system is comparable.

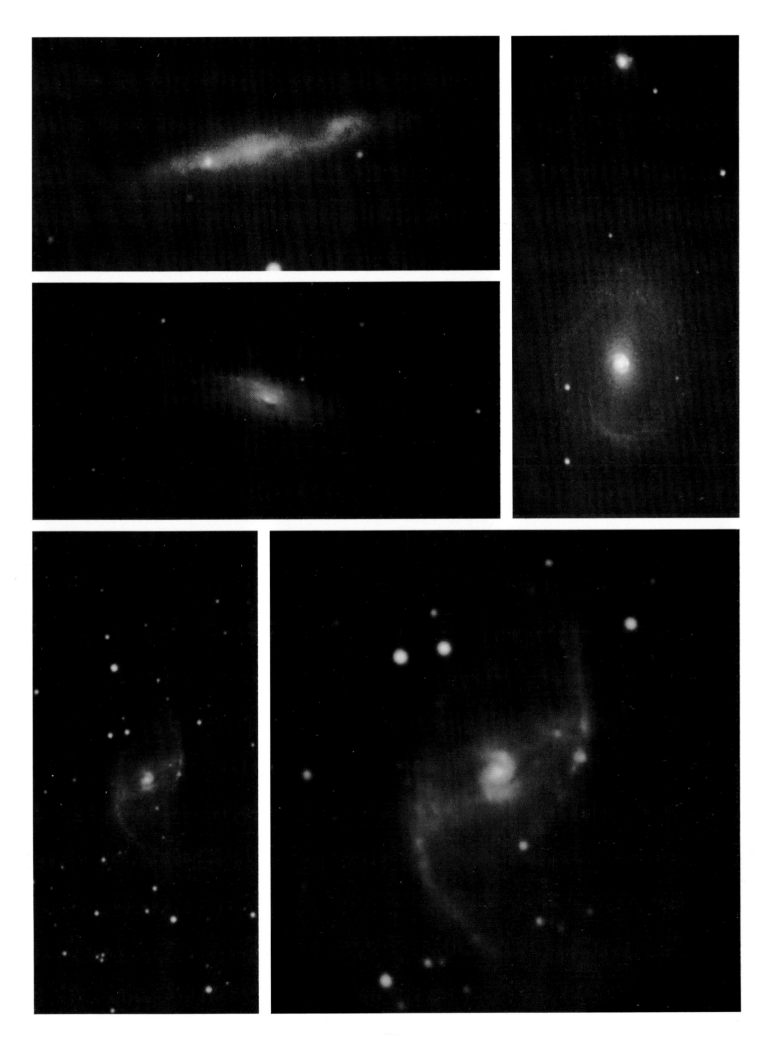

NGC 1532

1438 6 81 11 03 07.8   LC 1.0m
B=11.2   B−V=0.86   U−B=0.25   V=1019
.SBS2$/   Sab(s)I(Tides?)   T2 L4* N25

Dust lanes obscure and redden the nuclear region. The old population extends well into the surrounding disk, but it is interpenetrated by both intermediate aged and young stars over most of its extent. The spiral arms appear to have a large intermediate population, interspersed with a number of weak blue knots, with the exception of the one spiral arm segment in which a massive burst of star formation activity is occurring. The outlying arm seen stretching across the field below and to the right of the brighter main part of the galaxy contains no such luminous blue knots, nor does it appear to contain massive amounts of dust as is seen throughout the bright interior region. Instead, this arm is dominated by stars of an intermediate color, and hence an intermediate aged population.

The second galaxy in the field is NGC 1531. It appears to be composed mostly of yellow stars, although the brightest two regions, symmetrically located with respect to the center, are too bright to reveal any thing other than the dominant color. A number of small blue irregular dwarf galaxies, each one not much larger than a typical blue knot, are also present.

This print is a composite of two sets of exposures. The scale is enlarged by a factor of 1.7 times the standard enlargement factor used throughout the Atlas. The telescope was shifted slightly between the two sets of exposures. As a result, any visible defects appear as pairs. The two most prominent are at 29,2 and 34,2. Another is at 3,4. These particular defects are present in LC 1.0m observations from plate 1307 through plate 1440. See also NGC 1291 for example.

1533
1543
1549
1559
1560
1566
1569
1617
1618

NGC 1533

1436	6	81 10 31 07.2		LC 1.0m
B=11.8	B−V=0.96	U−B=0.52		V=560
.LB.−..	SB02(2)/SBa	T=3		N00

The yellow population is the only one visible.
The bar is distorted somewhat into a slightly
spiral appearance. Two blue stellar objects are
visible in and near the disk at 14,1 and 14,2. It
is not clear whether they are associated with
the galaxy or not.

NGC 1543

1408	6	81 11 01 08.1		LC 1.0m
B=11.5	B−V=0.97	U−B=0.47		V=1183
RLBS0..	RSB02/3(0)/a	T=−2		N00

Only an old yellow population is apparent.
The red flare is a reflection from a star outside
the field.

NGC 1549

1419	6	81 11 02 06.6		LC 1.0m
B=10.8	B−V=0.93	U−B=0.45		V=938
.E.0+..	E2	T=−5		N00

Again, an old yellow population is all that is
evident.
A blue stellar object is present at 03,4. The
dark spot is an artifact (see NGC 1291).

NGC 1559

1356	6	81 10 30 07.4		LC 1.0m
B=10.8	B−V=0.41	U−B=−.08		V=1025
.SBS6..	SBc(s)II.8	T=6		N00

The nuclear region and bar contain a mixed
populaion including young stars. Several violet
knots (high excitation blue knots) occur in the
arms. An exceptionally blue stellar object is
seen at 04,2. It is probably associated with the
galaxy.

NGC 1560

447	6	77 11 07 07.8		McD 2.1m
B=12.2	B−V=0.72	U−B=0.10		V=151
.SAS7./		T=7		N09

Individual red, yellow , green and blue stellar
appearing images are visible at about the same
brightness. This is the primary indication of
the detection of individual supergiant stars.
Refer to the discussion accompanying NGC
625. Compare with the stars in the NGC 604
field, which has a distance modulus of
approximately +25. From this comparison
one can infer a very approximate apparent
distance modulus for NGC 1560 of +26. Of
course there are many factors to consider, not
the least of which is the question of the
variance in brightness of supergiants, not only
as a whole, but also in systematic dependence
on the differences in the circumstances under
which they occur in different galaxies and even
in different places in the same galaxy.

NGC 1566

1409	6	81 11 01 08.4		LC 1.0m
B=10.2	B−V=0.85	U−B=0.00		V=1178
.SXS4..	Sc(s)I	T=4		N09

A classic example: (i) The nuclear region and
lens is dominated by the yellow old
population. (ii) The broad smooth zones of
the arms are dominated by the green
intermediate aged population. (iii) The
narrow spiral pattern is defined by the blue
knots, the new-born stellar population. As
density wave models predict, the dust lane,
seen as a shock wave compression zone and
the actual site of star formation, follows
behind the blue knots, which are drifting
ahead out of the dust in which they were born.
The Seyfert nucleus is overexposed.

NGC 1569

444	6	77 11 07 07.3		McD 2.1m
B=11.9	B−V=0.77	U−B=−.16		V=87
.IB.9..	SmIV	T=10	L=6*	N24

The central region is greatly overexposed, but
does not appear to be yellow. The outlying
region appears amorphous and green. This
galaxy is at very low galactic latitude, and is
reddened by dust in our own galaxy. If this
reddening is on the order of +0.3 magnitudes
in B−V then the smooth outer region would
have a color similar to the smooth regions in
NGC 4449 and NGC 4490. With the
exception of the blue object, the stellar
appearing objects in the galaxy are yellow and
red; and there are many such objects. If these
are normal supergiant stars, as the
overwhelming number of red objects suggests,
then, by comparison with the NGC 604 field
again, the apparent distance modulus must be
on the order of +26, and allowing for one
magnitude of extinction based on the evident
reddening the corrected distance modulus
would be approximately +25. This is
discrepant with respect to the currently
accepted distance modulus for this galaxy
which is greater than +28. At least a part of
this discrepancy is due to the difference in the
size of the regions from which the samples of
supergiants were selected; the restricted area of
the galaxy NGC 598 in which the NGC 604
field is located under-represents the brightest
supergiants in that galaxy, whereas the
brightest supergiants in NGC 1569 are more
completely represented. Accounting for
selection effects often turns a seemingly
straight-forward problem into one of
intractable complexity, the solution to which
frequently requires more or better data as is
certainly the case here.
Corwin's photometry of the bright star is
V=9.78, B−V=0.91, U−B=0.57.

NGC 1617

1386	6	81 10 31 07.7		LC 1.0m
B=11.2	B−V=0.95	U−B=0.42		V=778
.SBS1..	Sa(s)	T=1		N34

Although only the old yellow population of
stars is evident, dust is present in large
quantities, particularly in the arm to the lower
left of the nuclear region.

NGC 1618

954	3	79 11 08.6		McD 2.7m
B=13.3	B−V=0.80	U−B=−.05		V=5267
?.SBR3$.		T=3		N09

A faint arm containing several blue knots can
be seen extending above and to the right of the
nuclear region.

1622
1625
1635
1637
1642
1659
1667
1672
1688

NGC 1622

955	3	79 11 25 08.8			McD 2.7m

B=13.1   B−V=0.89   U−B=0.33

.SXR2*.   T=2 L=4 N32

A faint yellow appearing ring is visible.

NGC 1625

956	3	79 11 25 09.0		McD 2.7m

B=13.2   B−V=0.80   U−B=−.4   V=295

.SBT3*.   Sb/Sc   T=3*   N18

The nuclear region is flanked by two short straight arm segments, each containing some star formation activity. The outer disk is dominated by intermediate aged and young stars. A diffuse blue knot is visible at 01,1. A group of background galaxies is visible. The largest appears qualitatively similar to NGC 1625.

NGC 1635

1011	3	79 11 26 07.6		McD 2.7m

RSXR0..   T=0   N00

Compare with NGC 1512 and NGC 2545. A blue knot is seen in the faint spiral arm at 06,2. Another blue stellar object is present at 33,4.

NGC 1637

1010	3	79 11 26 07.4			McD 2.7m

B=11.6   B−V=0.65   U−B=0.05   V=626

.SXT5..   SBc(s)II.3   T=5   N09

The old population is well established in the bar. The ring is incomplete, and most of the pseudo-ring is actually well defined spiral structure, In a bar/arm/ring evolutionary scenario, this galaxy would follow NGC 1365, and would precede NGC 1241. NGC 1635, on the other hand, would follow NGC 1241. An exceptionally blue object is present at 34,3.

NGC 1642

957	3	79 11 25 09.1			McD 2.7m

B=13.2   B−V=0.75   U−B=0.00   V=4566

.SAT5*.   T=5   N00

The two red objects (35,1 and 30,3) appear to be background galaxies. The blue diffuse object at 24,2 is probably associated with NGC 1642.

NGC 1659

958	3	79 11 25 09.2		McD 2.7m

B=13.1   B−V=0.70   U−B=−.05   V=4444

.SAR4P. Sc(s)II−III   T=4 L=5 N00

The disk appears to have a significant population of intermediate age. Star formation activity is greater near the perimeter of the disk. Plume structure is evident in the disk.

NGC 1667

1012	3	79 11 26 07.7		McD 2.7m

B=12.7   B−V=0.70   U−B=0.04   V=4504

.SXR5..   Sc(r):IIp   T=5 L=3* N00

The bright ring in the inner disk appears white and patchy. This suggests blue star formation regions in the otherwise old yellow region. A bright blue knot is seen at 30,1. Note the background galaxy at 24,3.

NGC 1672

1420	6	81 11 02 06.9		LC 1.0m

B=10.3   B−V=0.62   U−B=0.00   V=1076

.SBS3..   Sb(rs)II   T=3 L=2 N15

Compare with NGC 613. Note in particular the faint suggestion of ring structure inward from the bright spiral arm segment ranging from 02,2 to 08,3. Note also that this spiral feature does not lie tangent to the end of the bar, but turns into the bar along the leading edge of the dust lane as in NGC 613. The brightest regions of star formation are near the two ends of the bar. Star formation is also visible in association with the bright nuclear region.

NGC 1688

1421	6	81 11 02 07.2		LC 1.0m

B=12.6*   B−V=0.50   U−B=−0.05   V=995

.SBS5..   SBc(s)II−III   T=5 N00

The nucleus appears white, and the bar as two discernibly green segments. See also NGC 600 and NGC 3319. This is one of a small class of very unusual systems.

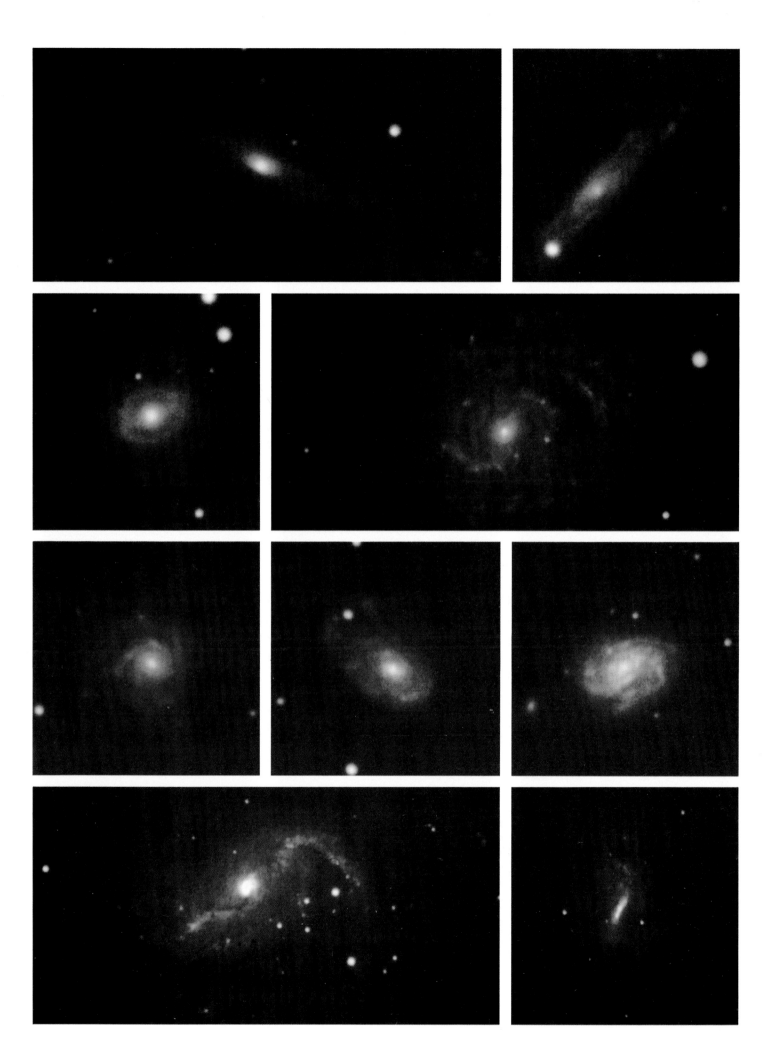

1700
1703
1720
1741
1744
1752
1779
1784

NGC 1700

959	3	79 11 25 09.3			McD	2.7m
B=11.9	B−V=0.95	U−B=0.49	V=3870			
.E.4... E3		T=−5	N32			

This galaxy appears to consist entirely of an old stellar population. Several background galaxies are visible in the field, one at 30,2 and another at 13,4. A blue object is at 11,4. Corwin's photometry of the bright star (09,7) is V=10.69, B−V=1 .17, U−B=1.17.

NGC 1703

1439	6	81 11 03 08.0		LC	1.0m
B=12.0	B−V=0.55	U−B=0.00			
.SBR3..		T=3	N24		

The brightest spiral region is superimposed on a larger disk of mixed population. This condition is unusual. Note the contrast with the otherwise somewhat similar galaxies NGC 3184 and NGC 3344. Plume structure is widespread in the outer disks of both NGC 1703 and NGC 3184.

NGC 1720

960	3	79 11 25 09.5		McD	2.7m
				V=4093	
.SBS2..		T=2	N14		

Star formation precedes a dust lane along the lower bar. Several blue stellar objects are present.
The blue flare is from a star outside the field.

NGC 1741

1018	3	79 11 26 08.6		McD	2.7m
B=13.9	B−V=0.38	U−B=−.55	V=3933		
.SBS9*/		T=9	N25		

NGC 1741A is the central object. It appears to consist of two separate galaxies, one corresponding to the brightest white knot with two short blue bar-like arms, and the other being the slightly fainter yellow-green system of diffuse patches immediately to the right and below the bright white knot. The blue knot at 04,1 could belong to either of these two galaxies. NGC 1741B is the galaxy at 00,3. The galaxy in the extreme lower right corner (11,8) is IC 399.
The blue flare is a reflection from a star outside the field.

NGC 1744

1357	6	81 10 30 07.7		LC	1.0m
B=11.7	B−V=0.49	U−B=−.15	V=579		
.SBS7..	SBcd(s)II−III	T=7	L=5*	N00	

This faint system appears to have a yellow nucleus and a number of blue knots.

NGC 1752

1013	3	79 11 26 07.9		McD	2.7m
B=13.2	B−V=0.87	U−B=0.25			
.SBR5*.		T=5	N18		

An old yellow population tends to dominate the inner disk. The partial ring appears intermediate in color. No luminous blue knots are visible.

NGC 1779

1016	3	79 11 26 08.3		McD	2.7m
B=12.9	B−V=0.90	U−B=0.40			
.SBR0$.		T=0	N00		

The nuclear region and ring are dominated by the yellow population, although some intermediate aged or young stars are present in the ring.

NGC 1784

1014	3	79 11 26 08.0		McD	2.7m
B=12.4	B−V=0.70		V=2182		
.SBR5..	SBbc(rs)I−II	T=5	L=3*	N32	

Although the bar is rather yellow, it is broken and patchy. Faint blue knots occur widely scattered in a large ring-like zone.

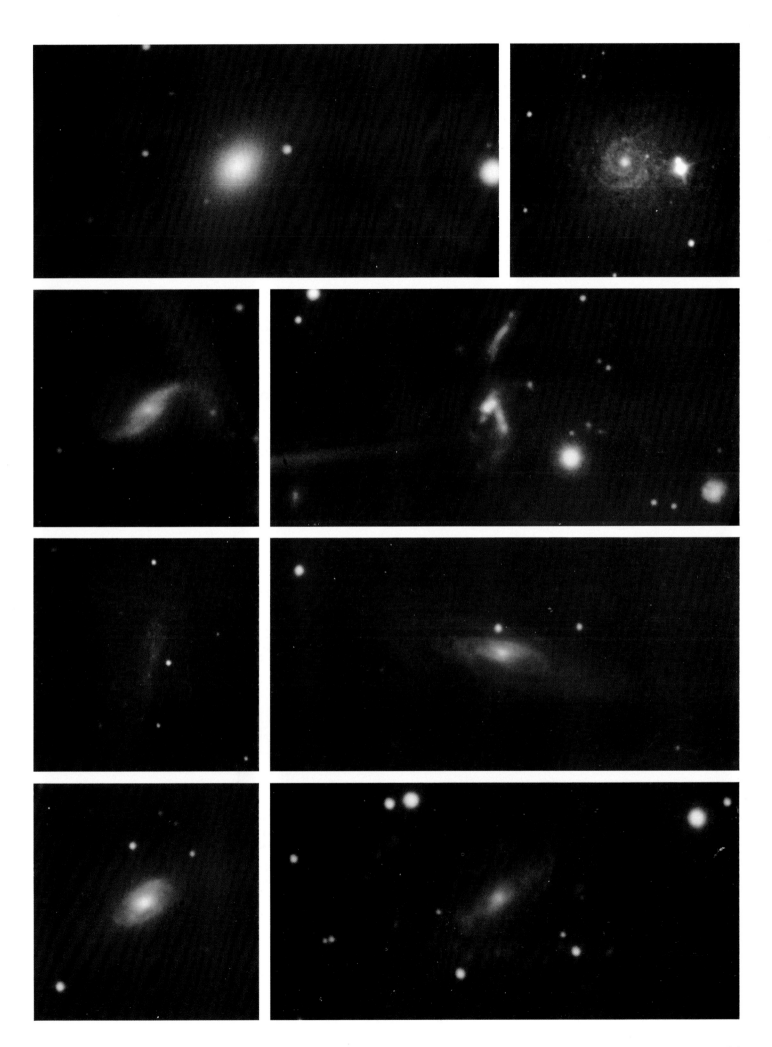

NGC 1792

1358 6 81 10 30 07.9			LC 1.0m
B=10.9	B−V=0.67	U−B=0.04	V=977
.SAT4..		T=4	N18

A small nuclear region is surrounded by a non-concentric multi-armed pattern of spiral features.

NGC 1808

1223 6 81 05 03 23.5			CTIO 0.9m
B=10.7	B−V=0.81	U−B=0.30	V=769
RSXS1..	Sbcp	T=1	N03

Dust lanes radiate from the central region at a steep inclination above and to the left of the nucleus. Below and to the right, however, such lanes are not evident, and the spiral pattern is not as steeply inclined.
A satellite crossed the field during the V exposure. A blue stellar object is visible at 30,7.

NGC 1832

1015 3 79 11 26 08.2			McD 2.7m
B=12.1	B−V=0.75	U−B=0.12	V=1760
.SBR4..	.SBb(r)I	T=4 L=3	N00

The spiral arms arc past the bar ends without noticeable perturbation. In this property NGC 1832 is somewhat similar to NGC 5921, but quite different from NGC 613, which is nevertheless similar to NGC 5921 in many ways. The brightest blue knots are in the outer disk region, associated with plumes rather than with the outer spiral arm.

NGC 1888

961 3 79 11 25 08.7			McD 2.7m
B=12.9	B−V=0.93	U−B=0.30	V=2353
.SBS5P.		T=5	N07

NGC 1888 and the elliptical galaxy NGC 1889 have similar red-shifts. Note the increase in the green stellar component with incresing distance from the center in NGC 1888 as is characteristic of spiral galaxies.

NGC 1954

962 3 79 11 25 09.8			McD 2.7m
B=12.5	B−V=0.58	U−B=0.00	V=2935
.SAT5..		T=5	N05

A narrow ring of faint blue knots surrounds the relatively isolated nuclear region.

NGC 1964

1224 6 81 05 03 23.8			CTIO 0.9m
B=11.5	B−V=0.78	U−B=0.22	V=1498
.SXS3..	SbI−II	T=3 L=3	N08

The disk consists of a bright inner region and a somewhat fainter outer region with a dark zone intermediate to the two. Star formation is occurring in both the inner and outer luminous regions.

NGC 2090

1225 6 81 05 04 00.1			CTIO 0.9m
B=11.4	B−V=0.67	U−B=0.22	V=1576
.SAT5..	Sc(s)II	T=5	N27

The disk is dominated by an old yellow population. A number of star formation regions are present.

NGC 2139

1425 6 81 11 02 08.1			LC 1.0m
B=12.0	B−V=0.34	U−B=−.30	V=1591
.SXT6..	SBc(s)II.3	T=6	N00

The nuclear region appears white. Star formation dominates the disk.

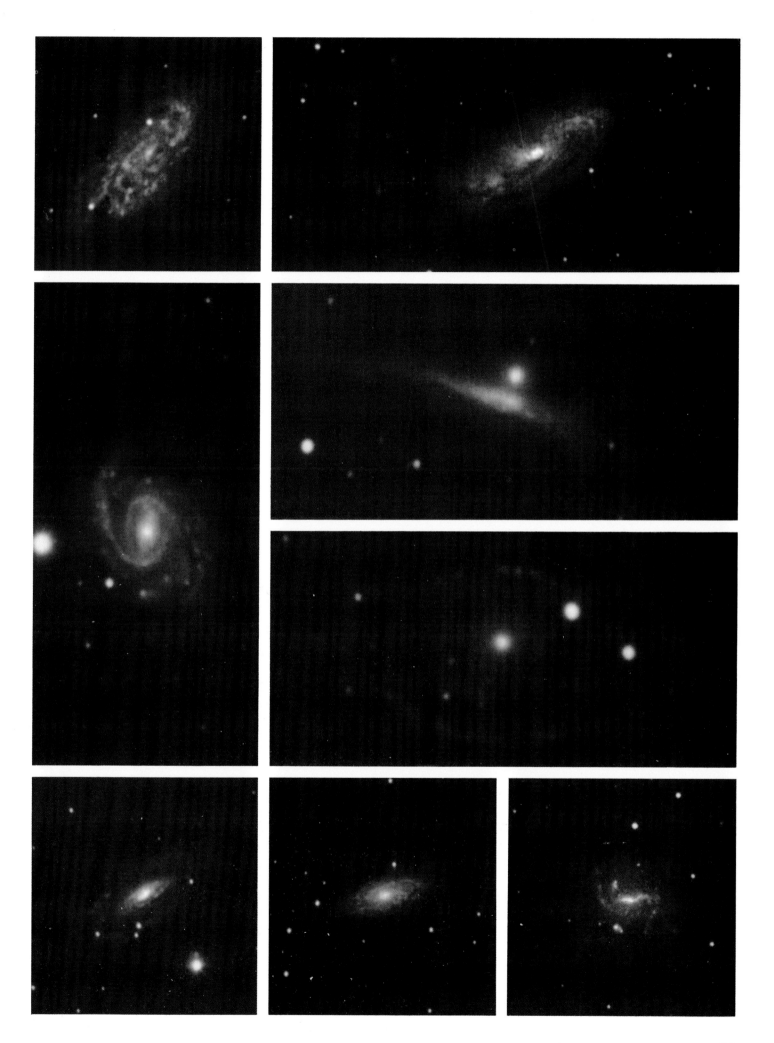

NGC 2146

1070	6	80 03 18 02.8			McD 2.1m
B=11.2	B−V=0.74	U−B=0.30		V=1028	
.SBS2P.	SbIIp		T=2		N00

The galaxy is strongly distorted. The
equatorial dust lane appears sharply tilted.
Note the considerable reddening due to dust.
Several blue knots are seen in the dust lane.
The main body of the galaxy is rather
amorphous and ranges in color from yellow to
green, indicating the presence of an old
population with at least a few stars of
intermediate age.

NGC 2188

1422	6	81 11 02 07.5			LC 1.0m
B=12.3	B−V=0.48	U−B=−.21		V=446	
.SBS9./	ScdIII		T=9		N00

A hint of green reveals the presence of stars of
at least intermediate age. There is no direct
evidence for a yellow population, although it is
not ruled out. See NGC 1313.

NGC 2196

1424	6	81 11 02 07.8			LC 1.0m
B=12.1	B−V=0.87	U−B=0.27		V=2080	
PSAS1..	Sab(s)I		T=1	L=2*	N00

The luminous central region is dominated by
an old yellow population, while the faint
narrow spiral arms exhibit the
characteristically blue regions of star
formation.

NGC 2217

1226	6	81 05 04 00.3			CTIO 0.9m
B=11.4	B−V=1.03	U−B=0.54		V=1243	
RLBT+..	SBa(s)		T=−1		N00

This galaxy is similar to NGC 1358.

NGC 2207

1440	6	81 11 03 08.3			LC 1.0m
B=11.3	B−V=0.70			V=246	
.SXT4P.	Sc(s)I.2		T=4	L=1*	N00

The two individual galaxies are readily
distinguished by their yellow nuclear regions.
Note also the differences in the disk
characteristics of the two galaxies. IC 2163, on
the left, appears here to be a background
object, partially obscured by NGC 2207. Note
in particular the dark lane crossing through the
central region of IC 2163 from below center to
upper right where it joins a bright spiral arm
segment in NGC 2207. This dust feature
appears to be an extension of that same spiral
arm. This circumstance offers a rare
opportunity to examine the distribution of
dust in a spiral arm such as this, and reveals
rather clearly that the dust distribution is much
more uniform than is the distribution of star
formation regions (as seen in the symmetrically
opposite arm). This condition is consistent
with the requirement that a critical density in
the dust-gas region must be reached before star
formation can begin to take place, and that
throughout much of the dust-gas arm the
density, although appreciable, is less than the
critical value. Of course, once initiated in a
given locale star, formation tends to maintain
itself through self-generation of local shock
compression sufficient to produce densities
well above the critical value for star formation.
In this regard see also NGC 6822.
Almost identical red-shifts for the two galaxies
(60km/s greater for IC 2163) establish these
two galaxies as an interacting pair.

NGC 2223

1227	6	81 05 04 00.5			CTIO 0.9m
B=12.1	B−V=0.77	U−B=0.31		V=2517	
.SXR3..	SBbc(r)I.3		T=3	L=3*	N00

A very low surface brightness galaxy with,
nevertheless, a distinct yellow nuclear region.

NGC 2256

1071	6	80 03 18 03.1			McD 2.1m
					V=5244
.LX.−*.			T=3		N00

The two white objects in the field are
foreground stars. A second, distant, galaxy is
located at 05,2.

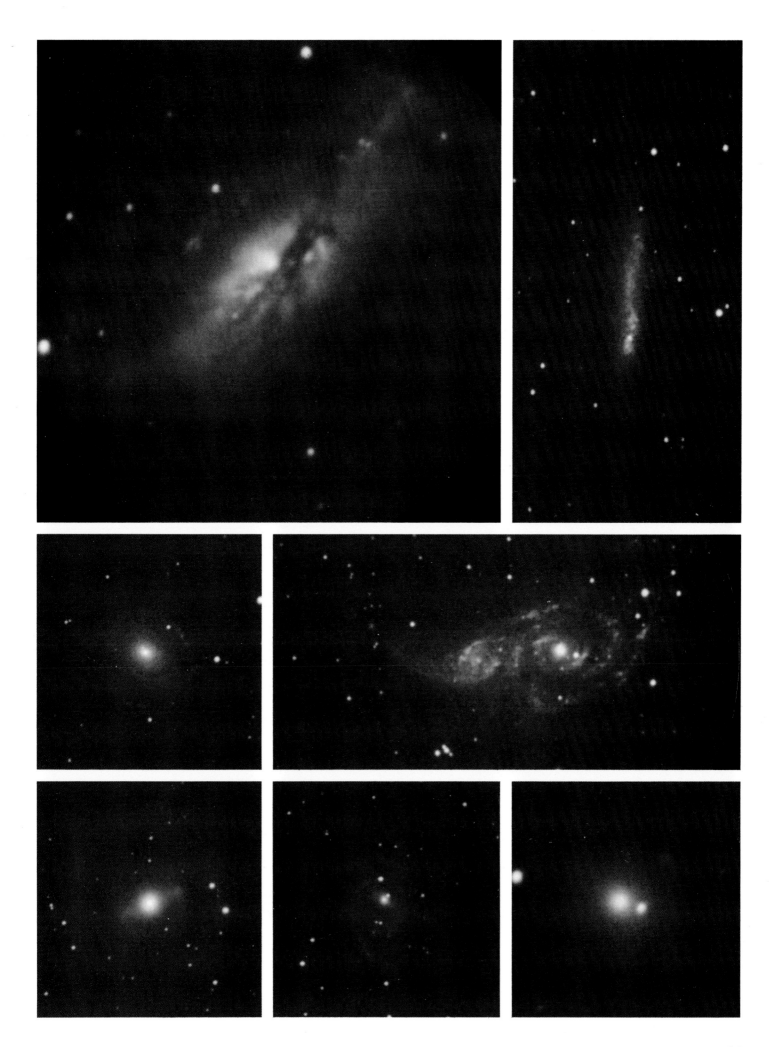

2280
2310
2336
2339
2347
2366
2389

NGC 2280

1209 6 81 05 01 23.5   CTIO 0.9m
B=11.1   B−V=0.55   U−B=0.15   V=1662
.SAS6..   Sc(s)I.2   T=6 L=3 N05

Very blue spiral arms wind smoothly around
the yellow nuclear region.
The field contains a number of blue stellar
objects.

NGC 2310

1210 6 81 05 01 23.7   CTIO 0.9m
B=11.6   B−V=0.60   U−B=0.25   V=917
.L..../ S02/3(8)   T=−2   N00

An old population system which appears also
to contain dust.

NGC 2336

479 6 78 01 09 04.6   McD 0.7m
B=11.1   B−V=0.66   U−B=0.05   V=2389
.SXR4..   SBbc(r)I   T=4 L=1 N09

Compare this 0.75 meter telescope photograph
with the 2.1 meter photograph below.
Several blue stellar objects are visible on the
right side of the field.

NGC 2339

964 3 79 11 25 10.2   McD 2.7m
B=12.3   B−V=0.74   U−B=0.10   V=2334
.SXT4..   SBc(s)II   T=4 L=4 N00

Multiple spiral arms appear to diverge from
the ring at irregular intervals.
The field contains many foreground stars.

NGC 2336

1024 3 79 11 26 09.6   McD 2.7m
B=11.1   B−V=0.66   U−B=0.05   V=2389
.SXR4..   SBbc(r)I   T=4 L=1 N09

Two bright blue knots punctuate the spiral
arm at the bottom of the field. The galaxy is
similar to, but lower in surface brightness
than, NGC 151.

NGC 2347

1026 3 79 11 26 08.0   McD 2.7m
B=13.3   B−V=0.77   U−B=0.13   V=4649
PSAR3*.   Sb(r)I−II   T=3   N00

The bright bar is tinted with the color of
intermediate aged or young stars.

NGC 2366

1025 3 79 11 26 09.8   McD 2.7m
B=11.4   B−V=0.40   U−B=−.35   V=252
.IBS9..   SBmIV−V   T=10 L=8 N13

This system is probably close to the threshold
of resolution into supergiants in this
photograph, but not enough red and yellow
stellar images are visible to permit certainty.
The most luminous blue knot (NGC 2363) is
so bright that it appears white. See NGC 604.
A red background galaxy is seen at 15,1.

NGC 2389

1051 6 80 03 17 03.7   McD 2.1m
B=13.3   B−V=0.50   U−B=−.12   V=3792
.SXT5..   T=5   N32

The system contains a relatively small bar in
comparison to the widely ranging spiral
structure.

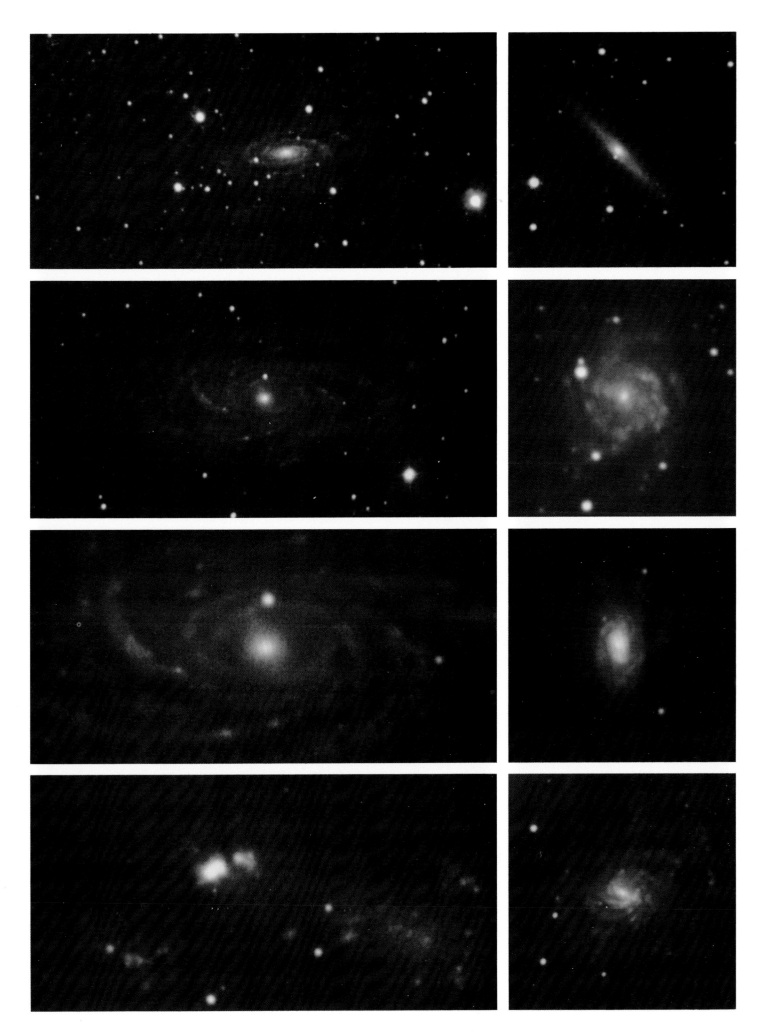

2403
2415
2417
2424
2427
2441
2442

NGC 2403

273b	6	77 10 15 11.3		McD 0.9m
B=8.8	B−V=0.50			V= 259
.SXS6..	Sc(s)III		T=6 L=5	N00

The old yellow population in the central region merges with an almost amorphous green intermediate population in the disk. Broad green arms define the disk spiral system, while beyond them the spiral pattern is mapped by the widely scattered blue knots. The most luminous blue knot is at 28,3.

A number of faint red, yellow and green stellar images, comparable in brightness to the two near 34,6, are seen throughout the disk. These appear to be supergiants at a distance modulus of about +27, in approximate agreement with the generally accepted value for this galaxy. See also NGC 625.

NGC 2415

1020	3	79 11 26 08.9		McD 2.7m
B=12.8	B−V=0.43	U−B=−.21	V=3779
.I..9$.			T=10	N03

An unusually high surface brightness galaxy. A range in surface brightness for galaxies of about this color index is seen in NGC 1313 (relatively low surface brightness), NGC 1309, and this galaxy.

NGC 2417

1388	6	81 10 31 08.2		LC 1.0m
B=12.9	B−V=0.75	U−B=0.10	V=2897
.SA.4*.			T=4	N00

Only the nucleus and several blue knots, including two close to the nucleus, are visible.

NGC 2424

1052	6	80 03 17 03.9		McD 2.1m
B=13.6	B−V=0.97	U−B=0.36	V=3307
.SBR3*/			T=3	N18

The perspective we have of this galaxy reveals the presence of massive amounts of dust, apparently in the outer disk. Several regions of star formation are visible.

A blue stellar object is present at 12,6.

NGC 2427

1211	6	81 05 02 00.0		CTIO 0.9m
B=12.3	B−V=0.78	U−B=−.25	V= 682
.SXS8..	Sc(s)II−III		T=8	N00

The field is at low galactic latitude, hence there are many foreground stars. A short yellow bar and a number of star formation regions are evident in NGC 2427.

A blue stellar object is visible at 15,1.

NGC 2441

1037	3	79 11 26 11.7		McD 2.7m
B=13.0	B−V=0.80			V=3782
.SXR3*.	Sc(r)I−II		T=3 L=3*	N00

Contrast the rather chaotic spiral pattern here with the smooth symmetric arms of NGC 2857. Each of these galaxies has a weakly defined bar and inner ring.

NGC 2442

1228	6	81 05 04 00.7		CTIO 0.9m
B=11.2	B−V=0.75	U−B=0.21	V= 384
.SBS3..	SBbc(rs)II		T=3 L=2	N27

This system appears to have massive dust lanes, particularly noticeable in the lower left arm. Most of the blue knots appear to be reddened.

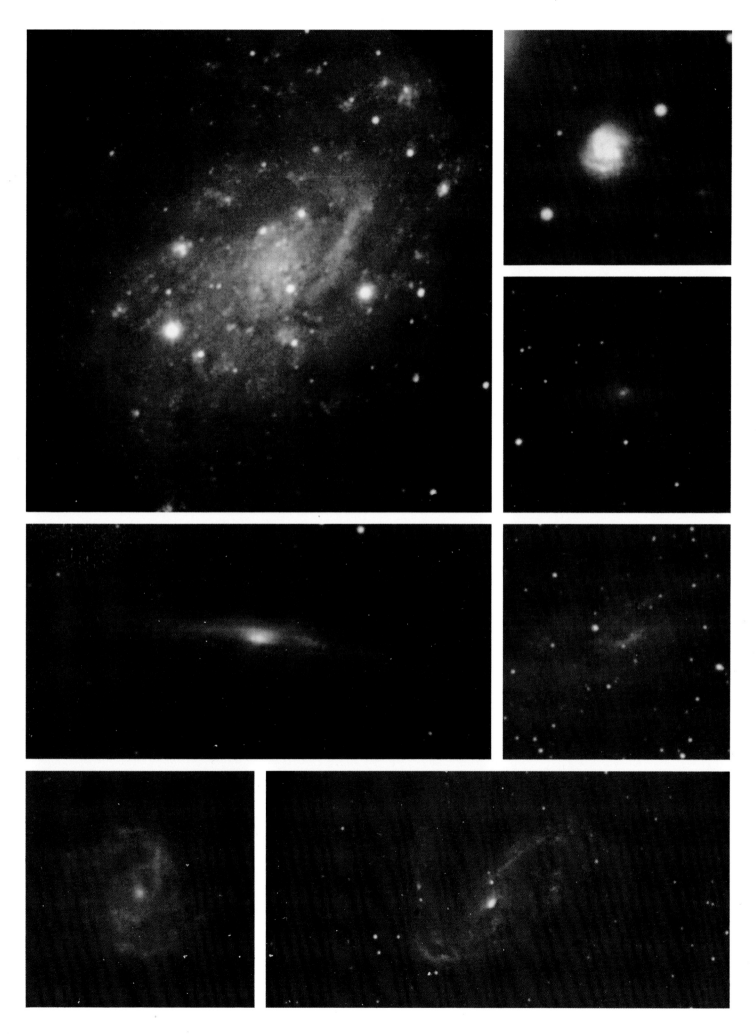

28

NGC 2445

1053	6	80	03	17	04.2		McD 2.1m
B=13.6		B−V=0.65		U−B=0.00		V=4035	
.RING.B					T=10		N09

The nucleus of NGC 2445 (27,1) remains approximately in the center of the pattern of blue knots despite the tidal interaction with the elliptical galaxy NGC 2444.

NGC 2500

1028	3	79	11	26	10.3		McD 2.7m
B=12.2		B−V=0.60		U−B=−.25		V=552	
.SBT7..		Sc(s)II.8			T=7	L=7	N00

Spiral structure is weakly organized, but star formation activity is widespread throughout the disk, apparently the result of a combination of both density wave and stochastic processes. An old population is visible in the nuclear region.
Corwin's photometry of the bright star is V=11.02, B−V=1.62 and U−B=1.88.

NGC 2523

1038	3	79	11	26	11.8		McD 2.7m
B=12.7		B−V=0.70		U−B=0.19		V=3608	
.SBS5*.		SBb(r)I			T=5		N33

The well established bar has weak characteristic longitudinal dust lanes. One arm bifurcates into two very narrow well defined arm segments. This condition indicates that processes supporting bifurcation are not necessarily synonymous with processes producing disruption and chaotic structure. On the other hand, the inclination of the outer segment is suggestive of a plume type structure.

NGC 2532

1112	4.5	80	03	21	03.7		McD 2.1m
B=13.0		B−V=0.62		U−B=−.05		V=5211	
.SXT5..					T=5		N09

The nuclear region has a well established old population. Although the inner arms are well defined and bright with star formation activity, the fainter part of the disk breaks up into widespread plume structure.

NGC 2460

1054	6	80	03	17	04.4		McD 2.1m
B=12.6		B−V=0.92		U−B=0.36		V=1543	
.SAS1..		Sab(s)			T=1	L=5*	N09

The bright central region is dominated by the old yellow population, as are the arms, although the latter exhibit a hint of the presence of stars of younger age.

NGC 2507

965	3	79	11	25	10.4		McD 2.7m
.S..0P.					T=0		N00

Compared to NGC 2460 the arms here are both greener and broader, suggesting a larger percentage of intermediate aged stars. The diffuse green object at 30,2 is probably a companion galaxy.

NGC 2535

966	3	79	11	25	10.6		McD 2.7m
B=13.1		B−V=0.52		U−B=−.10		V=4016	
.SAR5P.					T=5		N00

The white knot below right of the nucleus of NGC 2535 is probably a companion galaxy, as is NGC 2536 (below) as evidenced by both its proximity and the apparent tidal distortion of the intervening spiral arm of NGC 2535. The visible portion of the other arm terminates in what appears to be the bluest blue knot in the galaxy. NGC 2536 is undergoing a massive burst of star formation activity.

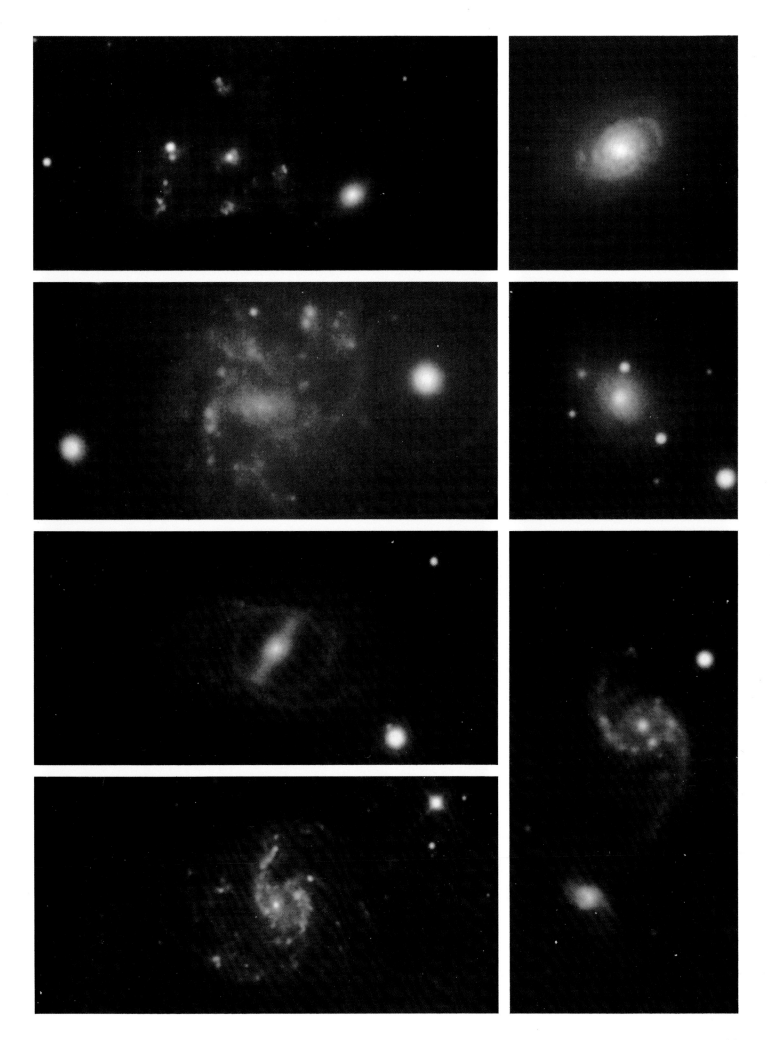

2537
2541
2543
2545
2549
2551

NGC 2537

1030	3	79 11 26	10.6		McD 2.7m
B=12.3	B−V=0.66	U−B=−.11			V=438
.IBS9P. ScIIIp			T=10		N30

This system is remarkable not only for its peculiar pattern of bright star formation regions, but also for its evident lack of a yellow nucleus despite the presence of a well developed disk of intermediate to old aged stars.

Corwin's photometry for the star is V=10.97, B−V=0.74 and U−B=0.34.

NGC 2541

1029	3	79 11 26	10.4		McD 2.7m
B=12.2	B−V=0.50	U−B=−.24			V=606
.SAS6..	Sc(s)III		T=6	L=7	N27

Two violet knots are present at 32,2. These could actually be individual O type stars in NGC 2541. This would put the supergiant branch at the threshold of detection, which appears possible, based on a number of exceedingly faint red images. If so, this galaxy would have a distance modulus of (very) approximately +29. See NGC 604 and NGC 625.

NGC 2543

1031	3	79 11 26	10.7	McD 2.7m
				V=2421
.SBS3..			T=3	N30

Longitudinal dust lanes are visible along the bar. The arms appear to cross the bar ends at a steep inclination, as in NGC 5921, but do not extend to form a bright ring as they do in that galaxy. In effect, the arms appear to attach to the bar at its leading edge, similar to the situation near 14,1 in NGC 5921.

NGC 2545

1057	6	80 03 17	05.2		McD 2.1m
B=13.2	B−V=0.80	U−B=0.17			V=3094
RSBR2..	SBc(r)I−II		T=2		N30

Similar to NGC 1635. The bar is relatively faint in comparison with the blue ring. Faint blue arms wind outside the ring.

NGC 2549

1034	3	79 11 26	11.2		McD 2.7m
B=12.0	B−V=0.94	U−B=0.55			V=1168
.LAR0./	S01/2(7)		T=−2		N30

What is evidently a background galaxy appears to be visible through the main body of NGC 2549 at 27,1. Another galaxy is visible at 06,3. Only an old population is seen in NGC 2549.

NGC 2551

1039	3	79 11 26	12.0		McD 2.7m
B=13.0	B−V=1.01				V=2375
.SAS0..	Sab(s)I		T=0	L=5*	N00

Faint green appearing spiral arms extend from the bright yellow lens.

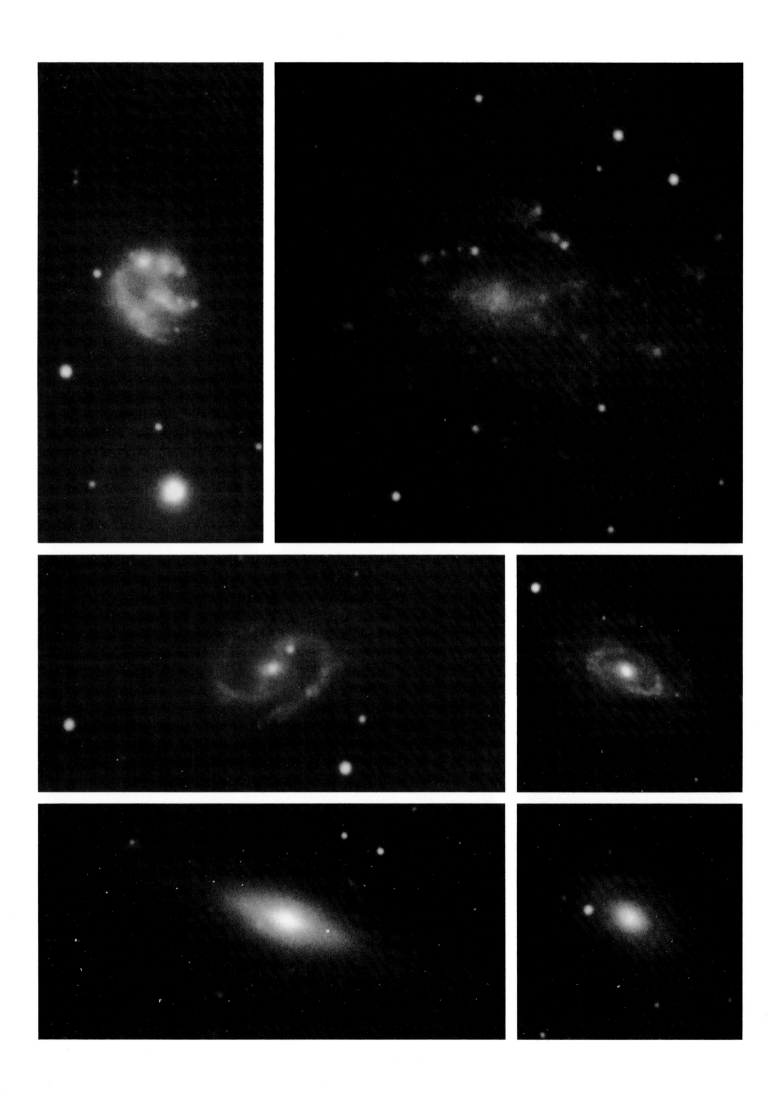

2552
2563
2595
2598
2608
2613
2614
2623
2633

NGC 2552

1033 3 79 11 26 11.0   McD 2.7m
B=12.8   B−V=0.59   U−B=−.20   V=557
.SAS9$.   Sc or SdIV   T=9 L=8 N33

There is no evident old population visible here, only a patchy amorphous green population and widespread blue knots. If there is an underlying old population this exposure is too short to reveal it. See also NGC 1313 and NGC 2188.

NGC 2563

968 3 79 11 25 10.9   McD 2.7m
B=13.4   B−V=1.02   U−B=0.61   V=4542
.L..0*.   T=2   N00

There is a faint cluster of galaxies in the direction of NGC 2563 which may account for the faint structural detail.

NGC 2595

969 3 79 11 25 11.0   McD 2.7m
B=12.6   B−V=0.65   U−B=0.08   V=4163
.SXR5..   T=5   N30

The bar is very faint, but appears to be yellow. There is a ring of star formation activity in the nuclear region, in the center of which the yellow nucleus is distinctly visible. Corwin's photometry of the star is V=9.01, B−V=0.78 and U−B=0.44.

NGC 2598

970 3 79 11 25 11.1   McD 2.7m
B=14.7   B−V=1.01   U−B=0.07
.SB.1$.   T=1   N30

The nuclear region appears yellow, but the disk exhibits the green color indicative of the presence of young and/or intermediate aged stars. The green stellar object is probably a foreground star.

NGC 2608

912 3 79 04 27 11.6   McD 2.7m
B=12.8   B−V=0.71   U−B=0.10   V=2053
.SBS3*.   Sbc(s)II   T=3 L=3* N00

The nucleus is bright. The bar, although not perfectly smooth, has yellow bar-end enhancements. The arms are active in star formation well past the bar ends, but not to the extent of a complete ring.

NGC 2613

1213 6 81 05 02 00.4   CTIO 0.9m
B=11.3   B−V=0.93   U−B=0.36   V=1444
.SAS3..   Sb(s)(II)   T=3 L=3 N00

The intricate structure of the dust lanes and the faintness of the blue knots combine to give the impression of a vast star system. The many foreground stars indicate that this galaxy is at low galactic latitude (b^{II}=10 degrees) and hence we expect several tenths of a magnitude increase in B−V and U−B due to dust in our own galaxy. Consequently the disk may actually be bluer than that of either NGC 2775 or NGC 2841.

NGC 2614

1040 3 79 11 26 12.2   McD 2.7m
                                     V=3614
.SAR5*.   T=5   N00

Two faint yellow bar-end enhancements are visible. Plume structure is pronounced between the spiral arms passing through 20,1 and 20,2.

NGC 2623

977 3 79 11 25 12.3   McD 2.7m
B=14.4   B−V=0.61   U−B=0.15   V=5355
.P.....   N04

The nuclear regions of apparently two individual galaxies comprise the central bright object. Faint arms extend to the upper left and lower right from the nucleus at the lower right. Other bright regions, one red and the other blue, appear to be associated with the other galaxy. The reddening may be due to dust in the arms of the first mentioned system.

NGC 2633

1076 6 80 03 18 04.8   McD 2.1m
B=12.8   B−V=0.97   V=2302
.SBS3..   SBb(s)I.3   T=3   N23

Spiral arms appear to attach at the bar ends, but prominent longitudinal dust lanes are not seen. Compare with NGC 210 which has a trace of star formation in rather broad bar-end regions, with NGC 600, and with NGC 6054 in which star formation is more extreme.

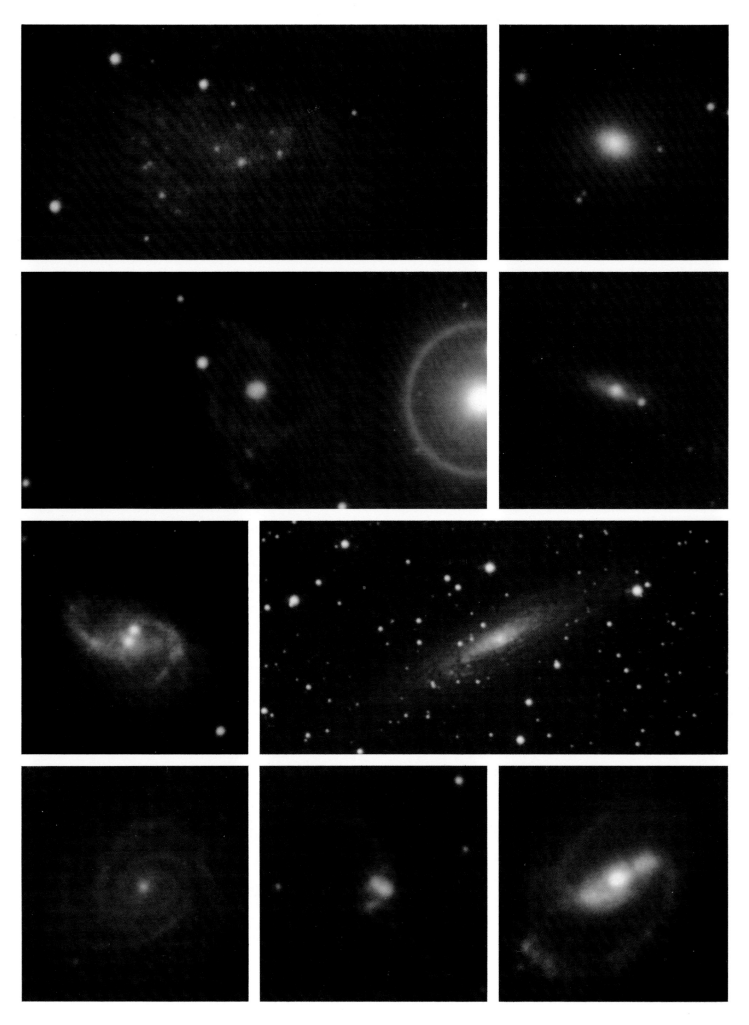

/>
2639
2642
2649
2654
2672
2673
2681
2683

NGC 2639

1035	3	79 11 26	11.4		McD 2.7m
B=12.6	B−V=0.89	U−B=0.37		V=3359	
RSAR1*$ Sa			T=1		N18

Patchy green regions in the disk reveal the presence of intermediate aged stars in this otherwise old population system.

NGC 2649

913	3	79 04 27	11.8	McD 2.7m
				V=3359
.SXT3*.			T=3	N00

The very symmetrical and regular spiral pattern here is similar to that of NGC 4321, but these galaxies are quite different in stellar population mix. Note the relatively green arms almost devoid of blue knots which dominate the arms of NGC 4321.

NGC 2681

763	6	79 03 25	03.3		McD 0.9m
B=11.1	B−V=0.80	U−B=0.32		V=760	
PSXT0..	Sa		T=0		N32

Two apparent arms cling closely to the nuclear region. Actually much of the appearance is due to the effect of dust. The arms exhibit a tint of green.

NGC 2672

971	3	79 11 25	11.3		McD 2.7m
B=12.6	B−V=0.98	U−B=0.57		V=4109	
.E.1+..	E2		T=−5		N00

NGC 2672 on the right and NGC 2673 on the left both appear to be 'normal' elliptical galaxies with old yellow stellar populations. There is no evidence for either young stars or dust.

NGC 2642

973	3	79 11 25	11.6		McD 2.7m
B=13.1	B−V=0.65	U−B=0.05		V=4226	
.SBR4..	SBb(rs)I−II		T=4	L=1	N32

This system is intermediate in morphological properties between NGC 2336 and NGC 2523.

Corwin's photometry of the bright star just off the edge of the field is V=9.38, B−V=0.56 and U−B=0.08.

NGC 2654

1036	3	79 11 26	11.5		McD 2.7m
B=12.7	B−V=0.95	U−B=0.54		V=1455	
.SB.2*/	Sab:		T=2	L=4	N17

This galaxy exhibits a 'box' shaped nuclear bulge. This configuration is evident in a number of galaxies which are viewed nearly edge-on. See for example: NGC 2683 (below), NGC 128, NGC 3628, NGC 4526 (tilted), NGC 4710, NGC 5529, NGC 5746, NGC 5854 (tilted), and NGC 7332 (tilted). The frequency of occurrence and the wide range of type (T) indicates that this phenomenon may derive from very general properties such as those giving rise to barred structure. A second point is that the stars involved are always the old yellow population. One possibility is that the dynamics involved may be slow to evolve, perhaps slower than the evolution of bar dynamics, since many bars are found which are dominated by intermediate and even young stellar populations. Another possibility is that the dynamics is initiated by a massive perturbation, and that the present appearance is the resultant harmonic oscillation. Perhaps NGC 520 represents an early stage of a 'box' structure before relaxation has had time to smooth out the dynamical irregularities. This explanation would require a prior encounter environment for all galaxies with this property, a condition which is subject to test. Perhaps the most significant factor, however, relates to the fact that a number of the box structures are tilted with respect to the major axis of the disk. Inspection of only the tilted systems reveals that the degree of binodality ranges from extreme in the case of NGC 128 to weak in the case of NGC 7332 to very weak in the case of NGC 4526 which is rather clearly a barred spiral. Next in the sequence is NGC 1023, a barred spiral with zero binodality. Hence we find a clear link between barred structure and binodality in which the latter would derive from out-of-plane orbital components for at least a sub-group of stars which comprise the bar.

NGC 2681

862	3	79 04 27	03.5		McD 2.7m
B=11.1	B−V=0.80	U−B=0.32		V=760	
PSXT0..	Sa		T=0		N32

Here the arms are seen to contain several bright bluish regions of star formation, notably at 01,1, in an old yellow stellar population. Note the faint, broad, almost bar-like extensions to the upper left and lower right visible in both photographs. No evidence is seen for star formation in these outer regions.

NGC 2683

462	6	78 01 08	07.0		McD 0.7m
B=10.5	B−V=0.89	U−B=0.29		V=242	
.SAT3..	Sb		T=3	L=4	N12

The yellow nuclear region has a slight tendency towards the box structure of NGC 2654. A number of blue knots are visible in the outer disk. Note the highly reddened region at 12,1. An inner dust lane is seen in the nuclear region.

A background galaxy is visible at 29,5.

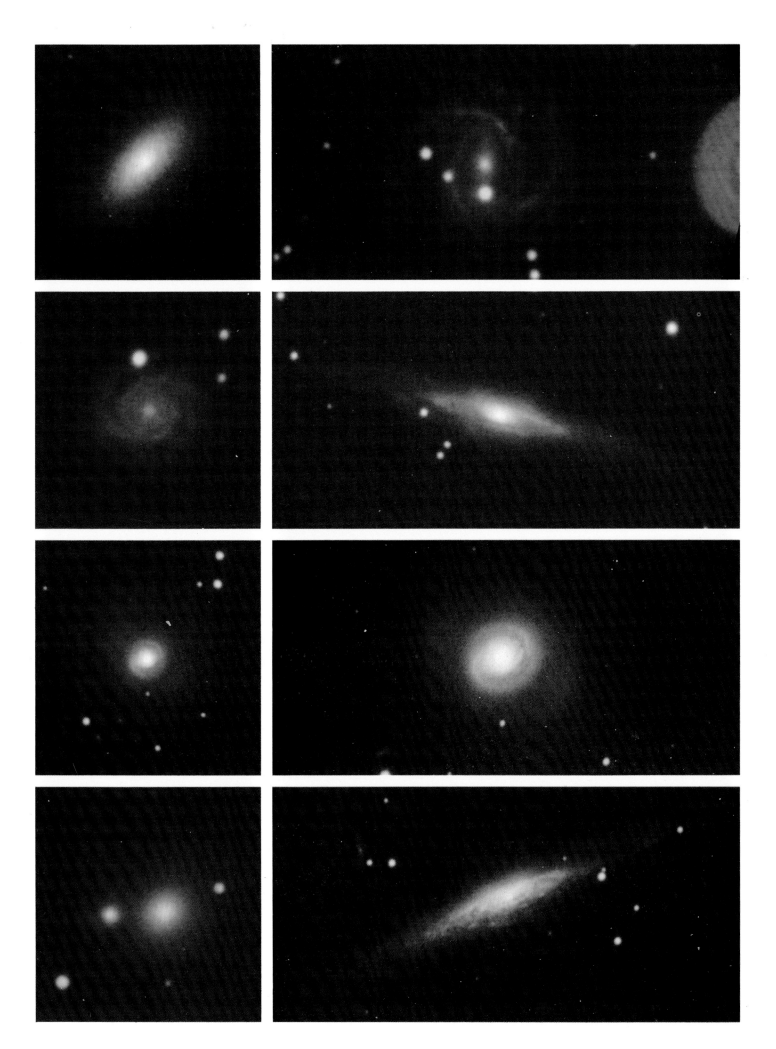

2685
2712
2713
2715
2742
2752
2768

NGC 2685

861 3 79 04 27 03.4   McD 2.7m
B=11.9   B−V=0.85   U−B=0.33   V=956
RLB..P.   S03(7)p   T=−2 L=4* N30

Many galaxies are peculiar, but few are as uniquely peculiar as NGC 2685. The swirl of dust lanes gives the appearance of a helix co-axial with a prolate spheroid of old yellow stars. Star formation regions are visible, notably at 33,2 to 02,2. Note the reddening and extinction due to dust to the lower left of center.

For an interpretation based on common morphological structures, replace 'prolate spheroid' with 'bar' and 'helix' with 'highly inclined disk' (presumably of a second, interacting, galaxy).

NGC 2712

914 3 79 04 27 11.9   McD 2.7m
B=12.7   B−V=0.68   U−B=0.09   V=1857
.SBR3*.   SBb(s)I   T=3 L=1 N00

A yellow bar is evident aligned along position angles 14 and 32. The arms are comprised primarily of young and intermediate aged stars. Note that the bright central region is more circular in projection than the disk.

NGC 2713

974 3 79 11 25 11.8   McD 2.7m
B=12.6   B−V=0.95   U−B=0.40   V=3688
.SBT2..   Sbc(s)I   T=2 L=3 N18

The bar is seen in a nearly end-on perspective. Light of the old yellow population dominates the system, although blue regions of star formation are visible in the very thin spiral features. A 'red knot' in the arm at 10,2 is either a foreground star or a background galaxy. Compare with NGC 210 and NGC 1358. See also NGC 6872.

NGC 2715

1514 6 82 03 28 02.5   McD 0.7m
B=11.9   B−V=0.57   U−B=−.12   V=1304
.SXT5..   Sc(s)II   T=5 L=4 N06

A yellow nuclear region and several blue knots in the disk are visible.

NGC 2742

512 6 78 03 05 06.4   McD 0.7m
B=12.3   B−V=0.62   U−B=0.03   V=1392
.SAS5*.   Sc(rs)II   T=5 L=3 N04

A relatively old and yellow disk extends from the nuclear region. Blue knots are embedded in the disk. These regions of star formation become most prominent near the edge of the older disk.
Corwin's photometry of the bright red star is V=7 .82, B−V=1.41 and U−B=1.67.

NGC 2752

972 3 79 11 25 11.5   McD 2.7m
                                          V=3904
.SB.3$/   T=3   N00

The old population dominates the disk of this system, although there is a suggestion of intermediate aged stars in the outer region of the disk.

NGC 2768

1077 6 80 03 18 05.2   McD 2.1m
B=10.9   B−V=0.93   U−B=0.55   V=1502
.E.6.*.   S01/2(6)   T=−5   N18

There is a suggestion of both reddening and extinction near or along the minor axis of this galaxy.

NGC 2770

```
1058  6  80 03 17 05.5              McD 2.1m
B=12.8    B−V=0.50   U−B=−.10    V=1911
.SAS5*.                     T=5        N05
```

The small nucleus is located in an apparent yellow bar. Star formation prevails throughout the outer disk.

NGC 2775

```
975  3  79 11 25 11.9              McD 2.7m
B=11.2    B−V=0.87   U−B=0.38    V=965
.SAR2..  Sa(r)               T=2        N00
```

The bright yellow inner disk is surrounded by a fainter yellow region. This outer yellow disk contains a multitude of embedded blue knots forming a broad ring of star formation regions with little if any global spiral structure. There appears to be evidence for a non-smooth structure in the outer yellow population. Compare with NGC 4736 and NGC 7217.

NGC 2776

```
513  6  78 03 05 06.8              McD 0.7m
B=12.2    B−V=0.56   U−B=−.05    V=2643
.SXT5..  Sc(s)I           T=5  L=3 N00
```

The spiral pattern is sharply defined by bright blue knots.
Reflections from a star outside the field occur in the upper left corner, with one extending to the lower right corner across the center of the galaxy.

NGC 2784

```
1263  6  81 05 07 00.4              LC 1.0m
B=11.2    B−V=1.15   U−B=0.72    V=435
.LAS0*.  S01(4)               T=−2       N00
```

The old population dominates this galaxy, although small patches of younger stars appear to be present.

NGC 2782

```
1059  6  80 03 17 05.8              McD 2.1m
B=12.1    B−V=0.66   U−B=0.00    V=2529
.SXT1P.                     T=1        N18
```

The bar of this galaxy is seen nearly end-on. Note the two dust lanes typical of many bars (e.g. NGC 1365). Reddening due to dust is seen in the lower lane, while the blue light of young stars is visible along the upper lane. Note also the presence of young blue stars in the nuclear region. A portion of ring structure at 27,2 is also very active in star formation.

NGC 2798

```
864  3  79 04 27 03.9              McD 2.7m
B=13.0    B−V=0.74   U−B=0.2     V=1709
.SBS1P.  SBa(s)(tides)      T=1        N12
```

The old yellow population of NGC 2798 (below) extends well into the disk in the form of broad smooth spiral arms. Regions of star formation occur along the leading edge of these arms and extend into the bright nuclear region. NGC 2799 (above) exhibits considerable star formation activity, although surprisingly little dust is evident in this edge-on view.

NGC 2787

```
1078  6  80 03 18 05.5              McD 2.1m
B=11.8    B−V=1.00   U−B=0.70    V=758
.LBR+..  SB0/a               T=−1       N35
```

This highly evolved barred galaxy exhibits long libration arcs which extend to form a ring. Overall this galaxy is very similar to NGC 936.

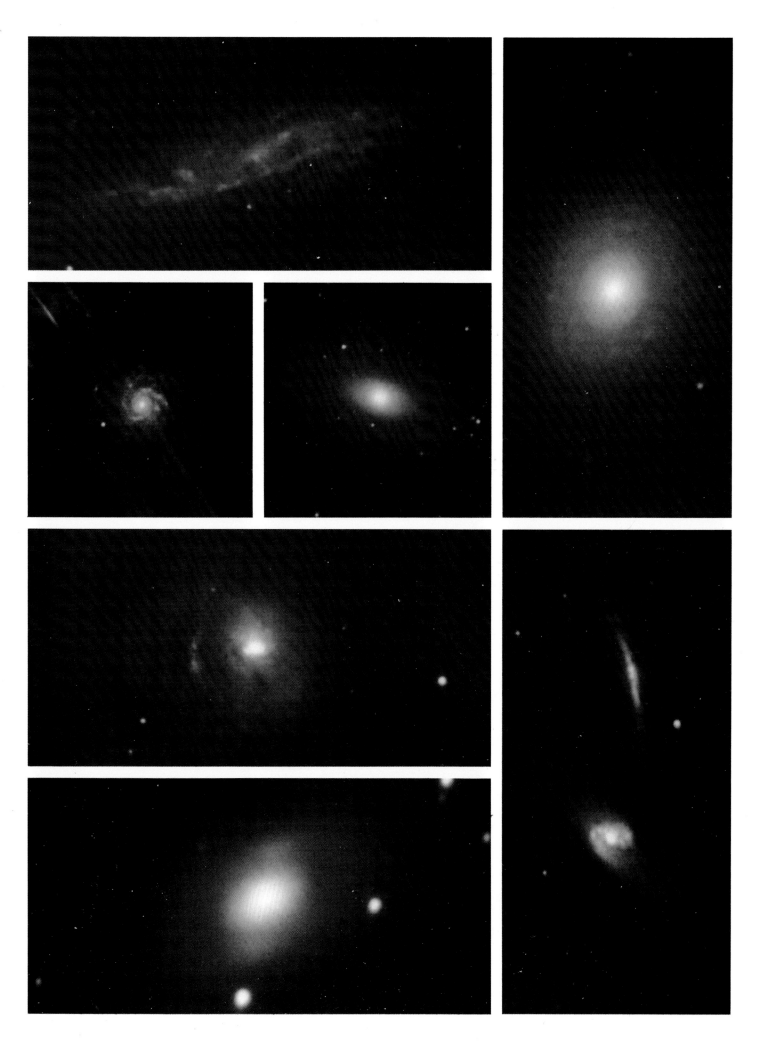

2835
2841
2848
2857
2859
2874
2872

NGC 2835

```
1214  6  81 05 02 00.6              CTIO 0.9m
B=10.8     B−V=0.47    U−B=−.15    V=617
.SBT5..    SBc(rs)I.2              T=5      N18
```

A mix of young and old populations comprise the spiral structure in this system. The inner arms are defined as well by the older yellow population as they are by the young blue knots.

NGC 2841

```
205  6  76 03 31 04.9               McD 0.9m
B=10.1     B−V=0.85    U−B=0.41    V=700
.SAR3*.  Sb                  T=3  L=1  N23
```

The broad outer disk is similar to that of NGC 2775, although long dust lanes are more in evidence here. These dust lanes reinforce the appearance of structure in the yellow population of the outer disk. The chaotic distribution of small blue knots makes it difficult to assess the amplitude and spatial frequency of surface density modulation of the yellow population component. Nevertheless, the distribution of the old population in the disk does not appear to be smooth; and the cause of this 'micro' turbulence in the old population becomes an interesting problem. Corwin's photometry of the star at 28,5 is V=11.05, B−V=0.58 and U−B=0.07.

NGC 2859

```
465  6  78 01 08 08.6               McD 0.7m
B=11.6     B−V=0.93    U−B=0.60    V=1657
RLBR+..    RSB02(3)            T=−1     N18
```

Compare with NGC 2787 and NGC 936. An exceedingly faint outer ring is visible in the original print. It is located about halfway from the center to the edge of the field.

NGC 2859

```
1115  4.5  80 03 21 04.5            McD 2.1m
B=11.6     B−V=0.93    U−B=0.60    V=1657
RLBR+..    RSB02(3)            T=−1     N18
```

The exposure for this 2.1 meter telescope photograph of NGC 2859 is reduced by a factor of 0.75 from the standard exposure used in the above photograph. The corresponding difference in apparent surface brightness is +0.3 magnitudes.

NGC 2848

```
1230  6  81 05 04 01.2              CTIO 0.9m
B=12.6     B−V=0.57    U−B=−.13    V=1922
.SXS5*.               T=5  L=7*  N00
```

The bright knot just to the right of the nucleus appears diffuse and is probably associated with the galaxy. Several blue stellar objects, some of which may be associated with the galaxy, are visible in the outer part of the field.

NGC 2857

```
863  3  79 04 27 03.7               McD 2.7m
                                    V=4919
.SAS5..               T=5      N00
```

Long narrow arms, defined by blue knots and stars of intermediate age, each extend more than 360 degrees in arc about the nucleus. There is a small blue inner ring. Faint long plumes connect the arms in the area of 12,2.

NGC 2874

```
976  3  79 11 25 12.1               McD 2.7m
B=12.8     B−V=0.80    U−B=0.22    V=3470
.SBR4..               T=4      N09
```

The yellow bar of NGC 2874 is seen nearly end-on. The inner ring structure is similarly foreshortened. The galaxy is similar to NGC 151, although the shape of the rings (in both cases) is actually elongated along the bar axis, as in NGC 5921. The disk appears to contain a complete mix of age populations. Star formation activity is prominent in the upper right arm as well as in the ring.
NGC 2872 (below) appears to contain a smooth yellow old population only. The other galaxy in the field is NGC 2873.

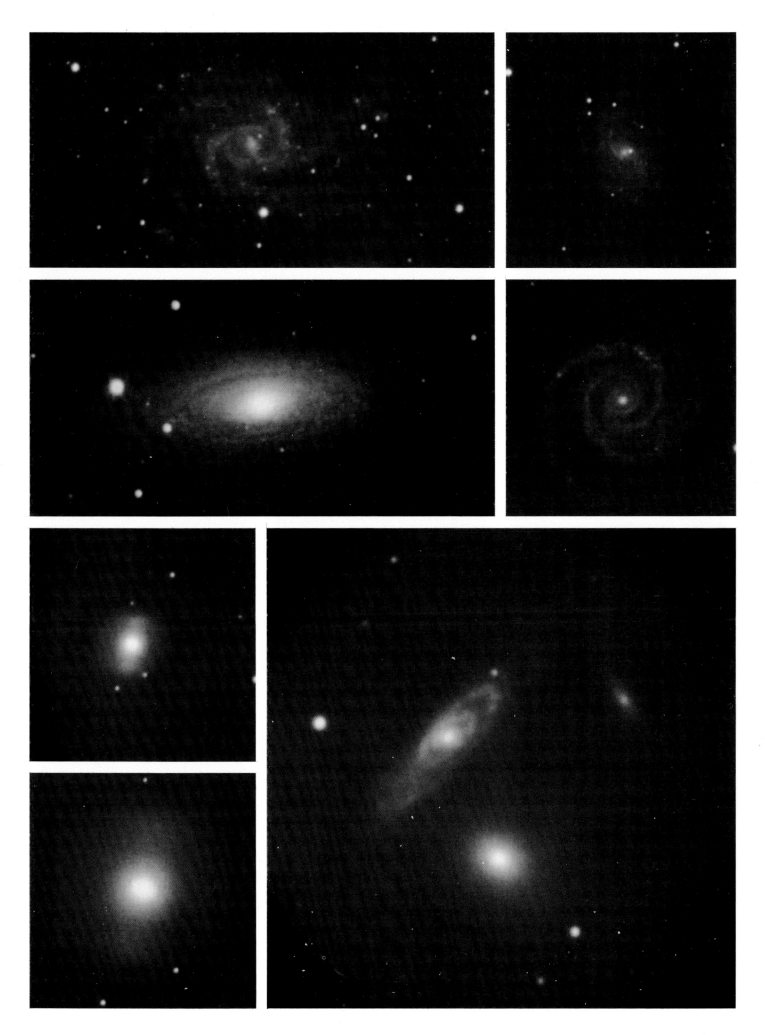

2903
2907
2935
2936
2937
2964
2967

NGC 2903

1504 6 82 03 27.			McD 0.7m
B=9.5	B−V=0.64	U−B=0.05	V=467
.SXT4..	Sc(s)I−II		T=4 L=2 N32

Compare this photograph taken with the 0.9 meter telescope with the photograph at the right taken with the 2.7 meter telescope. Note that although the color properties are rather similar an overall difference is discernible. This difference represents the typical variation in color representation throughout the atlas. Occasionally the deviation from the atlas norm may be considerably larger than this. Note, however, that prominent blues and yellows easily retain their identity although some of the intermediate colors become more difficult to judge. At the level of scientific application these properties establish in effect the useful working limits for interpretation of the atlas illustrations.

NGC 2903

1577 3 82 03 31 07.0			McD 2.7m
B=9.5	B−V=0.64	U−B=0.05	V=467
.SXT4..	Sc(s)I−II		T=4 L=2 N32

A distinct yellow bar crosses the disk of this galaxy, but because its structure is everywhere patchy and broken it is practically undetectable as a morphological entity (i.e. 'bar') without recourse to color information. If this is indeed a bar comprised of a genuinely old yellow population, the compact yellow knots would either have to remain intact over improbably long periods of time, or represent (improbable) reformations arising out of a dispersed distribution. If the effect were due to reddening by dust then the bar would be dark due to the corresponding extinction. A third and most likely possibility is that the yellow population distribution is in fact smooth, but embedded dust is creating the illusion of discrete knots of yellow stars. Somehow the illusion is a little too good for this to be entirely convincing, although it is clear that dust extinction is a major contributor to the appearance of the galaxy. Whatever its cause it seems appropriate to refer to the effect in the image as a 'color bar' because without the awareness of the color property the bar property is masked by the patchy structure which otherwise could be taken simply for bright regions of star formation in the spiral arms. In this sense NGC 2903 is unique in the atlas, and the cause of this aspect of its nature remains (or perhaps becomes) a mystery.

NGC 2907

202 6 76 03 31 04.2			McD 0.9m
B=12.7	B−V=1.10	U−B=0.55	V=1811
.SAS1$/	S03(6)p		T=1 N18

The remarkable feature of this galaxy is the equatorial dust lane which appears to extend far beyond the luminous disk of stars. Note the extreme reddening of the light from the nuclear region passing through this dust.

NGC 2935

1215 6 81 05 02 00.8			CTIO 0.9m
B=12.0*	B−V=0.72	U−B=0.30	V=1939
PSXS#..	SBb(s)I.2		T=3 L=1 N00

Thin arms beaded with blue knots wind out from a central bar. Note the clustering of blue knots at the ends of the bar in contrast to the linearity of their distribution along the arms.

NGC 2936

825 3 79 04 25 05.0			McD 2.7m
B=13.9	B−V=0.87	U−B=0.08	V=6794
.RING.B			T=10 N00

The yellow nucleus and distorted blue spiral features of NGC 2936 (center) are readily distinguished in our view of the cataclysmic interaction between it and NGC 2937 below. The appearance of the smooth arc of intermediate color, and hence of inferred intermediate age, suggests that this feature has an intrinsic existence or individual ontogenetic nature quite apart from any mere appearance arising out of density wave dynamics (in contrast with the spiral features of NGC 7743 for example). A shard-like patch of blue, probably a separate galaxy, is also visible near the bright star image at 02,4.

NGC 2964

865 3 79 04 27 04.0			McD 2.7m
B=12.0	B−V=0.71	U−B=0.00	V=1261
.SXR4*.	Sc(s)II.2		T=4 L=3* N14

This galaxy is very active in star formation, even into the bar. The bright nuclear region appears to have a spiral pattern (as in NGC 1365).

NGC 2967

743 6 79 03 24 04.8			McD 0.9m
B=12.3	B−V=0.66	U−B=0.08	V=1963
.SAS5..	Sc(rs)I−II		T=5 L=5* N00

The spiral pattern becomes prominent near the boundary of the yellow disk, similar in some respects to NGC 2649. There is no bar.

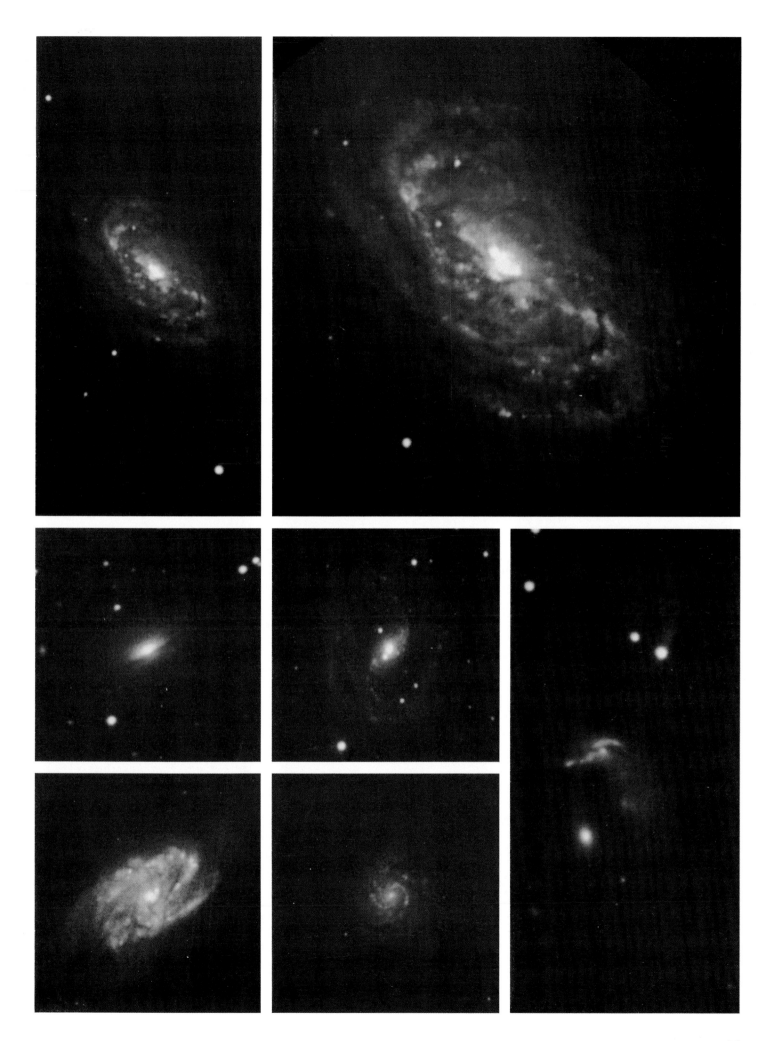

NGC 2976

1080	6	80 03 18 06.3		McD 2.1m
B=10.8 B−V=0.70				V=175
.SA.5P. SdIII−IV			T=5	N09

This remarkable galaxy may offer an opportunity to gain further insight into the nature of disk structures in galaxies. The nucleus is very small and faint. It seems probable that the nucleus is also very low in mass. If the mass to luminostiy ratio is constant for a given color population then there is virtually no central concentration of mass in this galaxy. As a result the entire disk rotates as a solid body. Hence the lack of spiral pattern may be equated to a lack of differential rotation in the disk. Nevertheless, there appear to be 'mass-cons' in the intermediate (yellow-green) population, and one can postulate the existence of a large scale pattern of weakly undulating density waves. The possibility of such waves notwithstanding, the most probable cause of the star formation activity seen here is in the realm of stochastic processes. In this regard it may be revealing that the brightest concentrations in the intermediate population are not coincident with the regions of maximum or even average star formation acitvity as indicated by the blue knots. On the contrary, the blue knots tend to be maximized near the perimeter of the disk. Note the distinct reddening in the dark lanes, indicating their nature as dust, and not the absence of stars.

NGC 2997

1216	6	81 05 02 01.1			CTIO 0.9m
B=10.0		B−V=0.73	U−B=0.15		V=805
.SXT5..		Sc(s)I.3		T=5 L=1	N00

Two dust lanes reach out from the nuclear region through the inner disk. The nuclear region also appears to be the site of star formation activity. Outside the old disk star formation activity, as evidenced by the blue knots, increases sharply, but is mostly constrained to the well developed spiral structure. The spiral arms appear to be a mix of intermediate aged and young stars.

NGC 2998

1060	6	80 03 17 06.1		McD 2.1m
B=12.7*				V=4769
.SXT5..		Sc(rs)I		T=5 L=3 N22

The dominant aspect of this galaxy is its system of spiral features. The nuclear region is definite but not pronounced. Several background galaxies are visible in the field.

NGC 3003

514	6	78 03 05 07.1			McD 0.7m
B=12.1		B−V=0.47	U−B=−0.20		V=1436
.S..4$.		Sc:3:		T=4 L=6*	N09

The spiral pattern is not well established in this galaxy, but the system is, nevertheless, dominated by the blue light of young stars. Note the blue stellar object at 01,6.

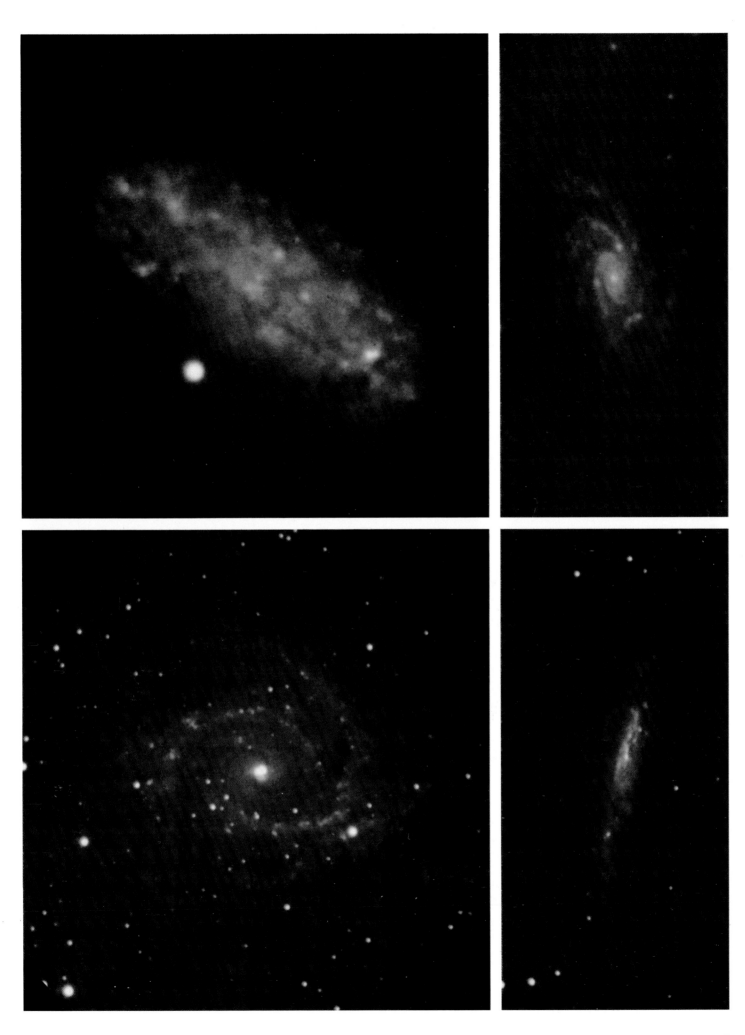

NGC 3031   M 81

463   6   78 01 08 07.6   McD 0.7m
B=7.7   B−V=0.93   U−B=0.48   V=95
.SAS2..   Sb(r)I–II   T=2 L=2 N09

Numerous dust lanes occur in the central region, yet no star formation is visible there. Many compact blue knots, regions of star formation, are seen in the outer disk. If an intermediate population is present in this galaxy it is too faint to be recorded in the standard exposure used here.

NGC 3032

723   6   79 03 22 05.3   McD 0.9m
B=12.6   B−V=.73   U−B=0.10   V=1507
.LXR0..   T=−2   N04

The inner bright zone appears white and may be the site of star formation, while the outer disk appears relatively smooth and yellow. See also NGC 278 and NGC 694.
Corwin's photometry of the bright star is V=9.05, B−V=0.57 and U−B=0.03.

NGC 3041

764   6   79 03 25 03.6   McD 0.9m
B=12.2   B−V=0.80   V=1294
.SXT5..   Sc(rs)II   T=5 L=4* N00

Faint spiral features with blue knots and intermediate aged stars form a somewhat disorganized spiral pattern about the small yellow nuclear region.

NGC 3034   M 82

464   6   78 01 08 08.0   McD 0.7m
B=9.3   B−V=0.87   U−B=0.30   V=388
.I.0../   Amorphous   T=0   N00

This object has been thought either to be exploding, or to have a massive concentration of luminous blue stars in its central region (invisible from our perspective due to the intervening dust). Light from the periphery of the object, particularly from the blue region visible below center, is strongly polarized. The cause of this polarization has been attributed to synchrotron radiation in the explosion model and scattering due to dust in the luminous blue core model. The system appears smooth except for the effects of dust which also reddens most of the galaxy. The smooth green region at the lower right indicates the presence of a large population of intermediate aged stars. No blue knots are visible. This suggests that either stars produced in the central region are being dispersed throughout the galaxy (dynamically improbable), or that a system-wide burst of star formation occurred in the intermediate past and ended then in at least the outer parts of the galaxy (not entirely unlikely). Thus the mystery of M 82.

NGC 3034   M 82

1156   1   80 04 19 05.1   McD 0.7m
B=9.3   B−V=0.87   U−B=0.30   V=388
.I.0../   Amorphous   T=0   N00

In this shorter exposure the extensive and complex pattern of dust is evident. The dust is causing extreme reddening as well as extinction over the entire bright area of the galaxy.

NGC 3054

1217   6   81 05 02 01.3   CTIO 0.9m
B=12.1*   B−V=0.75   U−B=0.25   V=1926
.SXR3..   T=3 L=4* N00

Spiral arms wind around a small bar. Enhancements in the arms occur beyond the bar ends, but approximately on line with it. These enhancements appear to be of intermediate age, suggesting the possiblility that the bar may be evolving slowly lengthwise.

NGC 3059

1229   6   81 05 04 01.0   CTIO 0.9m
B=11.8   B−V=0.65   U−B=−.15   V=954
.SBT4..   SBc(s)III   T=4   N00

The bar components are distinctly blue, indicating rapid star formation in the bar itself. This condition occurs infrequently, but is most common among galaxies of similar large scale structure.

NGC 3067

1160 6 80 04 19 08.2   McD 0.7m
B=12.7   B−V=0.68   U−B=0.04   V=1414
.SXS2$.   Sb(s)III   T=2 L=5 N18

The nuclear region is discernible by its yellow color. Several bright star formation regions are visible.

NGC 3079

466 6 78 01 08 09.0   McD 0.7m
B=11.2   B−V=0.64   U−B=−.05   V=1212
.SBS5./   Sc p: T=5 L=3* N09

Note the asymetrical distribution of yellow and blue populations along the minor axis. Some patchy structure is present in the yellow region.
Blue stellar objects are visible at 06,4 and 10,6. Corwin's photometry of the bright star just outside the field at 24,8 is V=7.94, B−V=0.73 and U−B=0.27; and for the star at 31,5 V=9.62, B−V=0.85 and U−B=0.51

NGC 3077

1578 3 82 03 31 07.3   McD 2.7m
B=10.6   B−V=0.80   U−B=0.15   V=148
.I.0.P.   Amorphous   T=0   N00

This object is unique in the atlas. Its color suggests a dust reddened complex of resolved high-excitation OB associations (such as NGC 604), surrounded by an older yellow population.

NGC 3109

1218 6 81 05 02 01.5   CTIO 0.9m
B=10.4*   V=131
.SBS9./ SmIV   T=9 L=8 N00

Individual red supergiants in this dwarf system are at the threshold of detection in this photograph. An uncorrected distance modulus of about +27 could be consistent with this appearance. See discussion accompanying NGC 625.

NGC 3113

1219 6 81 05 02 01.8   CTIO 0.9m
                                        V=811
.SAR7$.   T=7   N00

A faint galaxy comprising a typical mix of population ages.
Corwin's photometry of the star at the upper right is V=7.27, B−V=0.45 and U−B=0.00.

NGC 3115

744 6 79 03 24 05.1   McD 0.9m
B=10.0   B−V=0.95   U−B=0.57   V=476
.L..−./   S01(7)   T=−3   N00

An edge-on view of a highly evolved system in which we can see only old yellow stars. Dust is not evident.

NGC 3124

1266 6 81 05 07 01.3   LC 1.0m
B=12.7   B−V=0.70   U−B=0.00   V=3085
.SXT4..   SBbc(r)I   T=4 L=1* N00

An old populaion is evident in the nuclear region. Young to intermediate aged stars comprise the narrow spiral arms. The streaks are reflections from a star outside the field.

NGC 3147

1541 3 82 03 30 05.6   McD 2.7m
B=11.4   B−V=0.80   V=2881
.SAT4..   Sb(s)I.8   T=4 L=2* N00

A singular bright blue knot outshines the other major star producing centers. The yellow nuclear region dominates the galaxy as a whole. Note plume structure.

NGC 3162

515 6 78 03 05 07.5   McD 0.7m
B=12.1   B−V=0.60   V=1366
.SXT4..   Sbc(s)I.8   T=4 L=3* N00

This system is dominated by the blue of star formation regions and the green of intermediate aged stars in the spiral arms. The nuclear region appears white.

NGC 3166

1542 3 82 03 30 05.9   McD 2.7m
B=11.5   B−V=0.91   U−B=0.31   V=1203
.SXT0..   Sa(s)   T=0 L=5 N00

Thin yellow spiral features originate at each vertex of a four cornered bar. This quadrangular object is an extraordinary dynamical system.

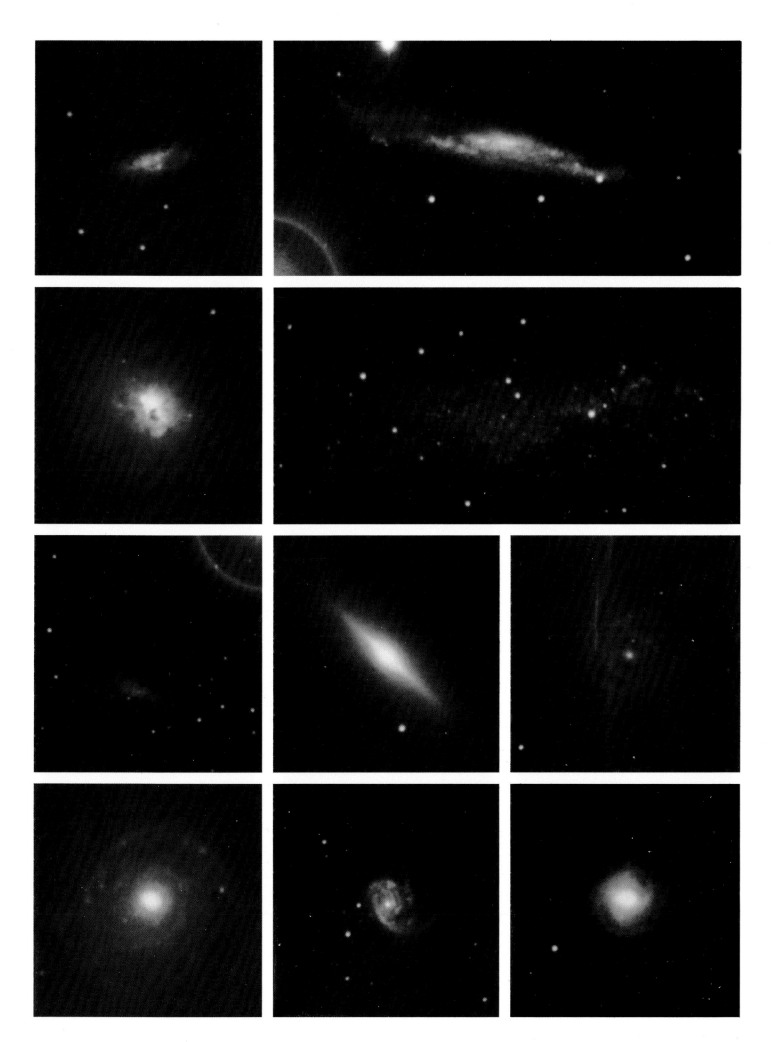

3175
3184
3185
3190
3198
3223

NGC 3175

1264 6 81 05 07 00.8			LC 1.0m
B=12.2	B−V=0.91	U−B=0.23	V=818
.SXS1$.	Sc(s)III:p	T=1	L=5 N05

The entire disk is dominated by an old population with star formation occurring in a relatively narrow zone intermediate between center and edge.

Corwin's photometry of the bright star is V=9.40, B−V=0.21 and U−B=0.11.

NGC 3184

208 6 76 03 31 05.3			McD 0.9m
B=10.4	B−V=0.65	U−B=−.02	V=593
.SXT6..	Sc(r)II.2	T=6	L=3 N00

The small nuclear region is yellow from the light of old stars. Beyond the nuclear region the spiral arms take over and dominate the disk of the galaxy. One arm is well defined by blue regions of star formation (blue knots). The other arm is better defined by the green population of intermediate age which pervades the disk. Many blue knots occur in the inter-arm region near 33,3.

Beyond the brightest spiral features, a number of plume structures are faintly visible. Those extending downward in the illustration from the arm at the lower left are dotted with blue knots, some of which form a curving alignment along the outer boundary of the plumes. It appears that such an alignment may represent the precursor of a future spiral feature. What is not so clear, however, is whether or not the plumes play a role in the development of such alignments or if they are simply by-products of other determinant processes.

NGC 3185

826 3 79 04 25 05.2			McD 2.7m
B=12.9	B−V=0.80		V=1147
RSBR1..	SBa(s)	T=1	L=5* N00

A blue ring surrounds the region of the bar. Two faint dust lanes are visible in the bar along position angles 05 and 24. The red spot at 27,1 is a defect.

NGC 3190

1061 6 80 03 17 06.4			McD 2.1m
B=11.9	B−V=0.97	U−B=0.45	V=1216
.SAS1P/	Sa	T=1	L=4* N04

The extraordinary pattern of dust lanes indicates a departure from equilibrium for the system as a whole. Consider the processes at work to constrain and maintain a continuous band of dust, such as the one rising to the upper left, over distances of galactic dimension.

Several blue stellar objects are seen in the disk. Perhaps this system of old stars is evolving towards a second generation of star formation as in NGC 4826.

NGC 3198

467 6 78 01 08 09.5			McD 0.7m
B=10.9	B−V=0.54	U−B=−0.05	V=691
.SBT5..	Sc(rs)I–II	T=5	L=3 N31

Contrast the appearance of this galaxy, overwhelmingly dominated by localized star formation acitvity, with the smooth relaxed distribution of old stars in NGC 3190. What forces have driven these two contemporaries at such radically different evolutionary rates?

NGC 3223

1220 6 81 05 02 02.0			CTIO 0.9m
B=11.8*	B−V=0.75	U−B=0.20	V=2574
.SAS3..	Sb(s)I–II	T=3	N00

A large nuclear region of old stars is surrounded by blue spiral features.

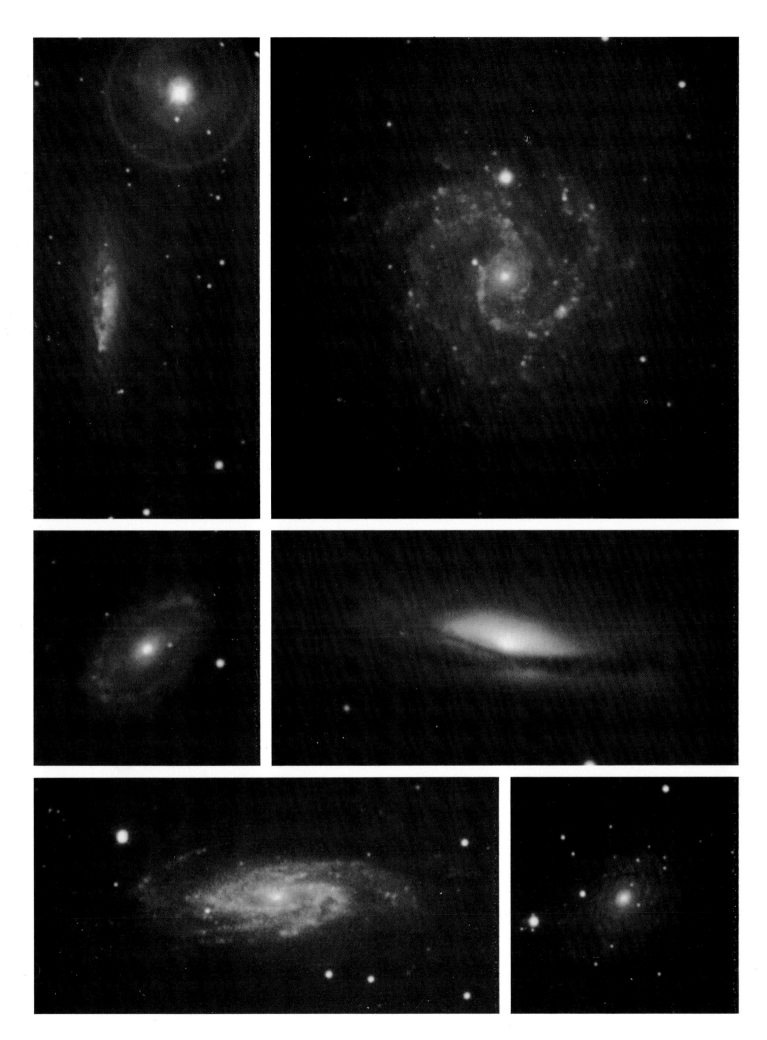

3226
3227
3254
3256
3261
3287
3299
3310
3319
3338

NGC 3226

765	6	79 03 25 03.8			McD 0.9m
B=12.3	B−V=0.93	U−B=0.45		V=1254
.E.2.*P S01(1)			T=−5		N09

NGC 3226 is the galaxy on the right. It appears to be a pure old yellow population system. NGC 3227 on the left has a blue disk which appears to lack pronounced spiral structure. It is somewhat similar to NGC 2976, although the radial brightness distribution has a much steeper gradient here.

NGC 3254

766	6	79 03 25 04.1			McD 0.9m
B=12.2	B−V=.66	U−B=−.05		V=1175
.SAS4..	Sb(s)II		T=4	L=3	N00

This galaxy has an extended faint disk which is blue, but (with the possible exception of a single blue stellar appearing object) lacks distinct blue knots.

NGC 3256

1265	6	81 05 07 01.0		LC 1.0m
B=11.6	B−V=0.62	U−B=−.10		V=2595
.P.....	Sb(s)p			N00

Star formation and dust occur in the central region of this chaotic system, while the outer region is mostly faint and yellow. Note the reddening in the dust lane at the immediate lower right of the nuclear region.

NGC 3261

1221	6	81 05 02 02.4		CTIO 0.9m
B=12.2*				V=2281
.SBT3..	SBbc(rs)I−II	T=3		N05

A barred system with long narrow spiral arm segments. The outer parts of both arms break up into plume structure. One arm appears to terminate at a bright blue knot.

NGC 3287

827	79 04 25 05.4		McD 2.7m
B=12.9*			V=1215
.SBS7..		T=7 L=7	N06

Star formation dominates this system. The bar and nucleus appear white.

NGC 3299

1116	4.5	80 03 21 04.8		McD 2.1m
				V=465
.SXS8..			T=8	N00

The system appears to be mostly yellow with a faint yellow nucleus. Images of stars of various colors are seen. There is little evidence for star formation, so the classification of these stars is very uncertain. Several objects appear to be diffuse, and may be star clusters. Despite the uncertainties, it is evident that this galaxy is relatively nearby.

NGC 3319

483	6	78 01 09 07.0			McD 0.7m
B=11.7	B−V=0.45	U−B=−.12		V=759
.SBT6..	SBc(s)II.4		T=6	L=3	N30

An excellent example of a blue bar. See also NGC 600 and NGC 1688. The spiral structure is irregular and open. Much of the light is from regions of star formation.

NGC 3310

741	6	79 03 24 04.0			McD 0.9m
B=11.2	B−V=0.32	U−B=−.45		V=1063
.SXR4P.	Sbc(r)p		T=4	L=3	N00

This is the most ultraviolet galaxy in the atlas according to photo-electric measurements (see data above). It is also one of the highest surface brightness galaxies illustrated. This galaxy is effectively one single mass of star formation activity.

NGC 3338

485	6	78 01 09 07.9			McD 0.7m
B=11.3	B−V=0.55	U−B=−.06		V=1191
.SAS5..	Sbc(s)I−II		T=5	L=3	N00

Broad blue spiral arms wind more than one full turn about the yellow nuclear region. Corwin's photometry of the bright star is V=8.94, B−V=0.28 and U−B=0.03

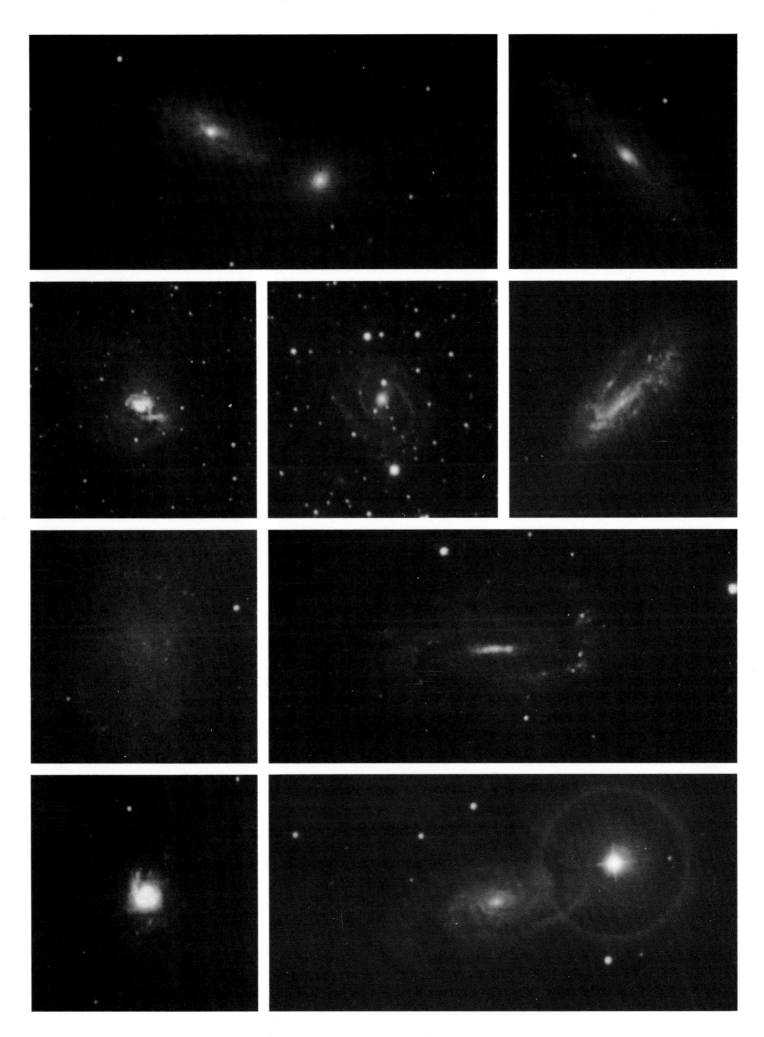

3344
3346
3347
3354
3351
3359

NGC 3344

468	6	78 01 08 09.9		McD 0.7m
B=10.5	B−V=0.55	U−B=−.10		V= 513
RSXR4..	SBbc(rs)I		T=4 L=3 N00	

A short yellow bar is visible. The inner ring structure appears to attach to or originate primarily at one end of the bar. The broad green components of the spiral arms have large intermediate age populations. Individual blue knots, sites of active star formation, are arrayed in narrow alignments along much of the spiral pattern. Plume structure is evident in the outer disk. Some plumes support blue knots. Active star formation is occurring beyond the perimeter of the disk visible here in blue knots at 00,4.

A distant cluster of galaxies is visible in the lower part of the photograph, and a blue stellar object is present in the same area at 20,9. Corwin's photometry of the bright star is V=9.76, B−V=0.65 and U−B=0.16.

NGC 3351   M 95

469	6	81 05 04 01.9		McD 0.7m
B=10.5	B−V=0.79	U−B=0.20		V=673
.SBR3..	SBb(r)II		T=3 L=3 N30	

The bar terminates inside the blue ring, similar to NGC 151. Dust lanes are visible in the bar. A yellow luminosity fills the disk interior to the ring. Spiral arms beyond the ring become faint and give rise to numerous faint plumes. Several blue stellar objects are visible in the vicinity of the spiral arms.

NGC 3351   M 95

1063	0.5	80 03 17 06.9		McD 2.1m
B=10.5	B−V=0.79	U−B=0.20		V=673
.SBR3..	SBb(r)II		T=3 L=3 N30	

A very short exposure with a larger telescope than the one above reveals a ring of blue knots surrounding a yellow nucleus. This is one of the group of active nuclear-ring galaxies. See also NGC 4314 and NGC 4321 among others. The red streaks are due to a star outside the field.

NGC 3359

484	6	78 01 09 07.5		McD 0.7m
B=11.0	B−V=0.55			V=1124
.SBT5..	SBc(s)I.8p		T=5 L=3 N09	

The spiral arms here are broad and diffuse in green light, although in some regions the alignment of blue knots within the arms is sharply defined. Star formation regions extend into the bar as well. Note that the left arm extends past the bar end (similar to NGC 151, and in contrast with NGC 613 for example). An irregular or disorganized ring is discernible.

NGC 3346

745	6	79 03 24 05.4		McD 0.9m
B=12.2*				V=1134
.SBT6..	SBc(rs)II.2		T=6 L=3 N00	

This faint barred system appears to have a blue nuclear region.

NGC 3347

1231	6	81 05 04 01.9		CTIO 0.9m
B=12.3	B−V=0.77	U−B=0.20		V=2642
.SBS5*/	SBb(r)I		T=5	N30

NGC 3347 (above) has a yellow inner region including nucleus, bar and faint ring. Outside this region very narrow blue spiral arms extend for considerable distances. One arm appears to end at a very bright blue knot.

The nearly white bar of NGC 3354 (below) is surrounded by a blue ring of active star formation. Note the somewhat similar shapes of the bright regions of these galaxies and the great contrast in their colors.

Several blue stellar objects are visible in the field.

3368
3377
3379
3395
3396
3423
3430

NGC 3368   M 96

486 6 78 01 09 08.3			McD 0.7m
B=10.1	B−V=0.86	U−B=0.27	V=773
.SXT2.. Sab(s)II		T=2	N00

The yellow disk extends slightly beyond the blue ring. Note the asymmetry in the visible dust distribution. It is not uncommon for dust on the near side of a disk to appear with greater contrast than dust on the far side, despite the otherwise general symmetry in the distribution of the smooth yellow population. The effect appears prominently in regions characterized by a rapid radial decrease in surface brightness of the smooth yellow component. The observed asymmetry would thus be consistent with a high ratio of back scattering to forward scattering by the dust (highly reflective opaque particles, for example). Such a property, combined with the difference in path lengths through the halo stars in front of the two dust regions (the greater path length to the far side disk results in a reduced contrast for the far side image in general) should account for the observed asymmetry.

NGC 3377

1064 6 80 03 17 07.1			McD 2.1m
B=11.1	B−V=0.84	U−B=0.31	V=450
.E.5+..	E6	T=−5	N09

A smooth yellow old population is all that is visible in this galaxy.

NGC 3379   M 105

768 6 79 03 25 04.7			McD 0.9m
B=10.2	B−V=0.94	U−B=0.52	V=756
.E.1...	E0	T=−5	N00

This elliptical galaxy is a photometric standard for surface brightness distribution. Note the uniformity of color (hue) over the entire range in luminosity from lightest to darkest perceptible. This degree of fidelity in color tracking is possible with the Dye Transfer color process used in making the color prints in this atlas. Because of this property color differences can be interpreted with satisfactory reliability at all discernible brightness levels in these photographs. This is a characteristic which is not afforded by most direct color photographic processes when applied to faint images such as galaxies.

NGC 3395

828 3 79 04 25 05.6			McD 2.7m
B=12.4	B−V=0.29	U−B=−.27	V=1595
.SXT6P*	Sc(s)II−III	T=6	N24

These galaxies (the lower one is NGC 3396) are two of the bluest (lowest B−V index) galaxies in the atlas. Their extreme nature makes their similarity all the more remarkable. Both are dominated by young stars with no compelling visible evidence for the presence of old stars. Perhaps these galaxies have originated in the relatively recent past as the result of gas stripping during the collision of other galaxies. It is also possible that they represent old systems which have evolved very slowly until the present encounter began, and as a result of this encounter they have both been shocked into a burst of rapid star formation which masks the presence of a weak older population.

NGC 3423

769 6 79 03 25 04.9			McD 0.9m
B=11.6	B−V=0.45		V=853
.SAS6..	Sc(s)II−III		T=6 L=3 N00

Rather low surface brightness with a distinct old population in the center and several active star formation regions in the outer disk characterize this galaxy.

NGC 3430

830 3 79 04 25 05.9			McD 2.7m
B=12.1	B−V=0.65	U−B=0.04	V=1529
.SXT5..	Sbc(r)I−II		T=5 L=3* N12

A very luminous blue knot punctuates this system. Note the bright blue knot coincident with the major bifurcation in the spiral arm at 30,1.

NGC 3432

1518	6	82 03 28 03.7			McD 0.7m
B=11.7		B−V=0.48			V=625
.SBS9./		Sc(II−III:)		T=9	N00

This very blue system shows no evidence in this photograph for an old yellow stellar population. The red flare is due to a star outside the field.

NGC 3486

211	6	76 03 31 06.5			McD 0.9m
B=10.8		B−V=0.52	U−B=−.17		V=674
.SXR5..		Sc(r)I−II		T=5	L=3 N00

The inner region is characterized by a bright yellow bar surrounded by a bright blue ring. Outside this ring the surface brightness in the disk drops precipitously, although blue knots are visible at a considerable distance from the center. Note in particular the bright blue stellar object at 13,5. The intermediate population is largely absent in this galaxy. Compare with the otherwise rather similar system NGC 3344.

NGC 3437

868	3	79 04 27 04.6		McD 2.7m
B=12.8*				V=1041
.SXT5*.		Sc(s)(III)	T=5	L=4 N34

The nuclear region is lost in the brightness of the disk, which is of mixed population. The diffuse yellow knot at 04,2 is not red enough to be a background elliptical, and is probably a small companion galaxy to NGC 3437.

NGC 3504

867	3	79 04 27 04.4			McD 2.7m
B=11.8		B−V=0.70	U−B=0.15		V=1479
RSXS2..		Sb(s)/SBbs(s)I−II		T=2	N00

The bar ends bend to conform to the spiral structure. Note the presence of young stars (implied by the blue knots) in the leading edge region of the bar just ahead of the bar dust lanes. Compare with NGC 7479. The disk is essentially defined by a form of plume structure which arises from both the leading and trailing edges of the bar. Contrast this system with NGC 151 and note the similarity with NGC 613. The distribution of the yellow population (excepting the effects due to dust extinction) is remarkably similar to NGC 7743.

Note the patch of blue knots beyond the bar/arm interface at 03,3. These blue knots are strongly aligned in an outward or radial direction. Compare with the region of the blue arm at the upper left in the NGC 625 photograph. It seems evident from this latter comparison that bar-end morphology defined by brightness structure may change radically, possibly even oscillate, over relatively short time scales in comparison with the time scale for evolution of the bar itself. Very faint diffuse outer spiral arms trail out from the regions of blue knots at the outer bar ends.

NGC 3511

1292	6	81 05 08 02.1			LC 1.0m
B=11.5		B−V=0.58	U−B=−.13		V=976
.SAS5..		Sc(s)II.8		T=5	L=3* N27

A very broad spiral pattern is visible. The arms are patchy but contain few blue knots. Compare with NGC 3227. There is a sequence of arm prominence in similar patchy disk dominated systems beginning with NGC 2976 (none) to NGC 3227 (very slight) to NGC 3511 (moderate). See also NGC 3917 and NGC 7314.

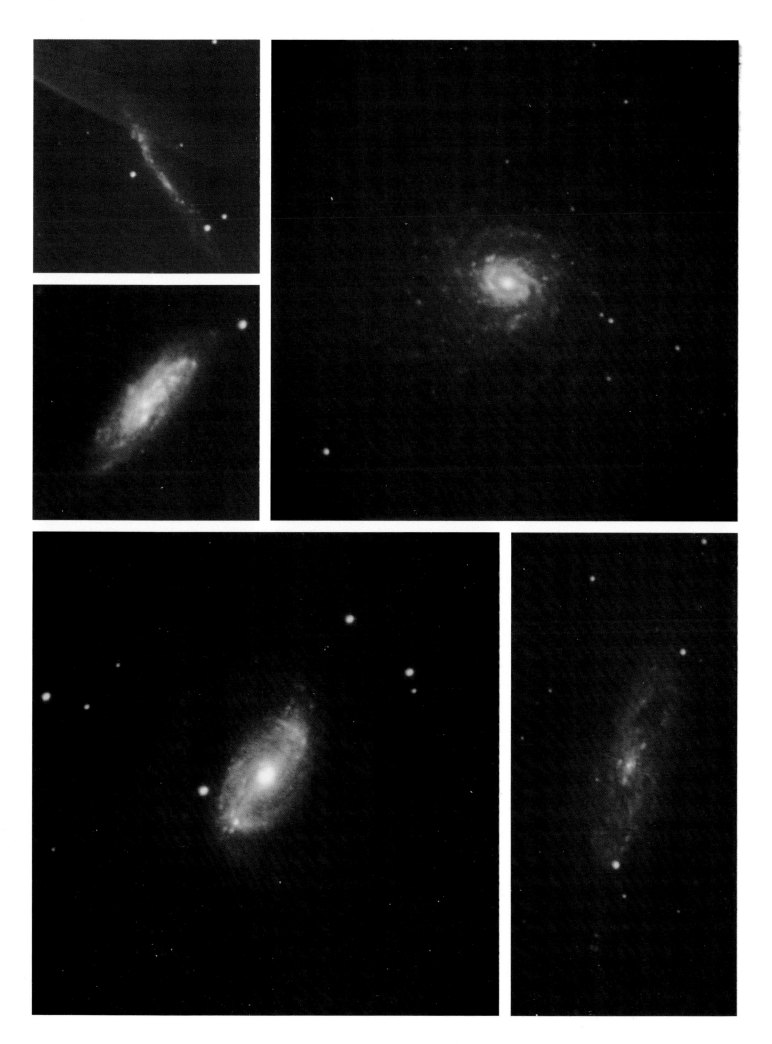

44

3521
3549
3556
3561
3583
3596
3614

NGC 3521

770 6 79 03 25 05.2   McD 0.9m
B= 9.7   B−V=0.84   U−B=0.15   V=640
.SXT4..   Sb(s)II−III   T=4 L=3 N09

Here we see a large intermediate age population in a disk comprised of numerous short spiral features each in turn comprised of plumes. The nuclear region is pronounced. Several blue knots are visible in the disk.

NGC 3549

1125 6 80 03 21 06.6   McD 2.1m
B=12.7   B−V=0.60   U−B=0.2   V=2942
.SAS5*.   Sbc(s)II   T=5 L=4 N28

The nuclear region is rather small, but the yellow disk is extensive. Beyond the inner disk long arcs of contiguous blue knots mixed with intermediate aged stars are visible. Plumes are seen radiating from the vicinity of these arcs. A singular bright blue knot is present at 06,4.

NGC 3556

241 6 76 04 01 04.7   McD 0.9m
B=10.6   B−V=0.61   U−B=−.01   V=772
.SBS6./   Sc(s)III   T=6   N18

The inner disk is yellow although there is no visible nuclear region *per se*. Star formation regions and massive chaotic dust clouds vie for dominion over the outer disk region. Spiral structure is not evident. There is an impression of large vertical excursions of dust lanes from the principal plane.
The blue spot at 30,1 is a defect.

NGC 3583

1126 6 80 03 21 06.9   McD 2.1m
B=12.2*   V=2213
.SBS3..   Sb(s)II   T=3 L=3 N19

A finely structured yellow population dominates this galaxy. Some blue knots are evident; others are lost in the glare of the yellow light. Some of the structure is due to dust and some is due to embedded blue knots, but there remains a residual fine structure in the yellow population which is not easily understood, and which may represent a color balance error of about 0.2 in B−V and U−B. Photoelectric measurements of this system would help to resolve this question.

NGC 3561

866 3 79 04 27 04.3   McD 2.7m
                                         V=8760
.L..0*P   T=−2   N14

This remarkable group of galaxies exhibits a number of curious features. Perhaps the strangest aspect of the interaction visible here is the sharply defined jet emanating from NGC 3561b in a direction opposite to NGC 3561a and terminating in an extraordinarily luminous blue knot. NGC 3561a exhibits the blue-bar phenomenon more strongly than any galaxy in the atlas with the exception of NGC 6054 which (incredibly) appears in the field of another galaxy with a long thin jet terminating in blue knots. The outer disk is distorted and faint blue regions are visible on the original print extending down and to the right from NGC 3561a. It seems somehow unlikely that the energy exchange resulting in the jet emanating from NGC 3561b can be understood simply in terms of gravitational tidal interaction.
The galaxy at the top of the field appears green, although at its distance individual blue knots would not be resolved but would blend in color with the intermediate and old stellar populations. The galaxy at 01,4 is largely dominated by an old stellar population and appears distorted.

NGC 3596

795 6 79 03 26 06.6   McD 0.9m
B=11.6*   V=1030
.SXT5..   Sc(r)II.2   T=5 L=3 N00

Compare this system with NGC 3184.

NGC 3614

796 6 79 03 26 06.9   McD 0.9m
B=12.2*   V=2334
.SBS9..   Sc(r)I   T=9   N00

Blue arms and a yellow nucleus are visible in this faint galaxy.

45

NGC 3621

1507	6	82 03 27 06.9	McD 0.7m
B=10.0*			V=455
.SAS7..	Sc(s)II.8		T=7 L=6 N22

This galaxy is similar to NGC 3733. The blue stellar object at 18,2 is probably associated with the galaxy. Plume structure is widespread.

NGC 3623   M 65

470	6	78 01 08 11.1	McD 0.7m
B=10.1	B−V=0.90	U−B=0.41	V=666
.SXT1..	Sa(s)II		T=1 L=3* N25

The smooth old yellow population dominates the entire luminous disk of the galaxy. A dense dust lane transforms into a line of star formation regions including a very bright blue knot at 08,2.

NGC 3626

1121	3	80 03 21 05.5	McD 2.1m
B=11.7	B−V=0.83	U−B=0.35	V=1363
RLAT+..	Sa		T=−1 L=3* N09

As in NGC 3623 the smooth old yellow population is widespread. A ring of dust surrounds the inner disk, but is mostly visible on the near side of the galaxy (see discussion of NGC 3368). Star formation regions are visible in the dust lane. See also NGC 4826. One must question whether an edge-on view of this galaxy might appear somewhat similar to NGC 4710.

NGC 3627   M 66

1579	3	82 03 31 07.4	McD 2.7m
B= 9.7	B−V=0.70	U−B=0.22	V=583
.SXS3..	Sb(s)II.2		T=3 L=3* N26

The yellow bar is finely structured, as in NGC 3583. Here most of the structure is evidently due to dust. Green plume structures, some with blue knots, radiate from the bar boundary. The overall color balance here is between 0.1 and 0.2 magnitudes too red.

NGC 3631

746	6	79 03 24 05.7	McD 0.9m
B=11.0	B−V=0.60		V=1245
.SAS5..	Sc(s)I−II		T=5 L=1 N00

The nuclear region is bright, but the surrounding disk lacks the smoothly distributed older population seen in otherwise similar galaxies such as NGC 628. Note the weak broad intermediate age component in the spiral arms, and the widespread plume structure.

NGC 3628

517	6	78 03 05 09.5	McD 0.7m
B=10.1	B−V=0.80		V=728
.S..3P/	Sbc		T=3 N00

A box structured nuclear region is evident. Note the broadening of the dust lane towards the right. Some faint blue knots are seen embedded in the dust.

NGC 3642

773	6	79 03 25 06.1	McD 0.9m
B=11.6	B−V=0.47		V=1728
.SAR4*.	Sb(r)I		T=4 L=1 N00

Only the yellow nuclear region and several very faint blue knots are bright enough to record in this photograph.

3664
3665
3672
3675
3686
3705
3717

NGC 3664

872 3 79 04 27 05.2   McD 2.7m
B=13.3   B−V=0.40   U−B=−.25   V=1241
.SB.9P$ SBmIII−IV   T=9   N00

The greenish color in part of the bar hints at
the presence of a few yellow stars, but it is also
possible that this galaxy contains no old stars
at all.

NGC 3665

1122 3 80 03 21 05.7   McD 2.1m
B=11.7   B−V=0.94   U−B=0.53   V=2012
.LAS0.. S0₃(3)   T=−2   N09

Here we see a galaxy apparently composed
entirely of old stars. The dust lane appears
relatively smooth and there is no distinct
indication of star formation in this
photograph.

NGC 3675

774 6 79 03 25 06.4   McD 0.9m
B=10.9   V=735
.SAS3.. Sb(r)II   T=3   L=3 N32

Broad greenish circular arms surround the
nuclear region. No individual blue knots are
visible. There is a dense inner dust lane. This
galaxy could be evolving towards a system
similar to NGC 3626.

NGC 3672

1543 3 82 03 30 06.1   McD 2.7m
B=11.7*   V=1737
.SAS5.. Sc(s)I−II   T=5   N27

The yellow nuclear region is surrounded by an
old population disk. Several blue knots are
visible in the arms.

NGC 3686

1065 6 80 03 17 07.5   McD 2.1m
B=12.0   B−V=0.58   V=931
.SBS4.. SBbc(s)I−II   T=4 L=3 N02

The spiral arms are broad and definite, but
highly disorganized at the local scale. Plume
structure is widespread, but not strongly
organized either. The blue knots for the most
part do not follow clear spiral delineations.

NGC 3705

1544 3 82 03 30 06.3   McD 2.7m
B=11.6   B−V=0.80   U−B=0.08   V=891
.SXR2.. Sab(r)I−II   T=2   N04

The perspective appears to be that of a bar
seen end-on. One bar dust lane is visible below
right of the bright nuclear region. The ring
contains regions of star formation.
This and all other 1500 series photographs are
approximately 0.1 to 0.2 magnitudes too red
due to an apparent drop in the UV sensitivity
of the image tube which occurred sometime
prior to the final observing run.

NGC 3717

1267 6 81 05 07 01.6   LC 1.0m
B=11.9   B−V=0.95   U−B=0.48   V=1481
.SA.3*/ Sb(s)   T=3   L=5 N11

Apparently this system could be called a dust
disk galaxy. Dust obscures and deeply reddens
the entire near half of the galaxy. The visible
yellow stars are probably mostly halo stars
(above the principal plane), and these together
with the back scattering function of the dust
determine the appearance (or lack) of the dust
disk on the far side of the galaxy.

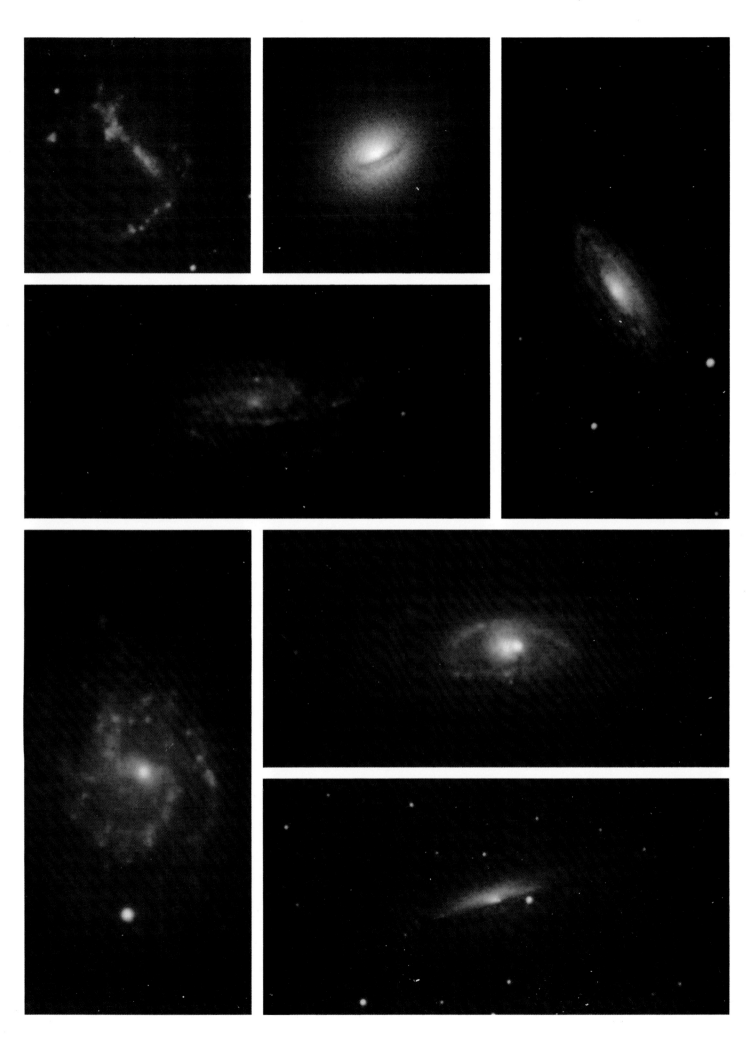

3718
3726
3733
3745
3746
3748
3750
3753
3754
3756
3780

NGC 3718

```
488  6  78 01 09 09.4          McD 0.7m
B=11.2     B−V=0.73    U−B=0.16     V=1095
.SBS1P.   Sap              T=1        N00
```

The remarkable feature of this galaxy is the
curving dust lane that cuts almost exactly
across the center of the galaxy. If this dust
represents a disk, then the disk either is
distorted, or does not pass through the center
of the galaxy, or both. Note both the greenish
color of stellar population and the extreme
reddening due to the dust.

NGC 3726

```
244  6  76 04 01 05.1          McD 0.9m
B=10.9     B−V=0.51              V=818
.SXR5..   Sc(r)I−II        T=5  L=2  N27
```

The short bar is surrounded by a broad patchy
ring with small scale structure similar to that
of NGC 3511. Outside this ring is a single
short spiral arm together with numerous
plumes containing most of the active star
formation regions in the galaxy.

NGC 3733

```
1127  6  80 03 21 07.2          McD 2.1m
B=12.8     B−V=0.52    U−B=−.20     V=1278
.SXS5*.                    T=5        N09
```

A small yellow nucleus is visible in this low
surface brightness galaxy. Note the similarity
with NGC 3511 with the exception that here
the radial decrease in surface brightness is
much steeper than in NGC 3511. The galaxy is
also quite similar to NGC 3621. Several
outlying blue knots are visible.
The bright star at the lower right is a probable
supernova.

NGC 3745

```
870  3  79 04 27 04.9          McD 2.7m
B=15.5     B−V=0.88    U−B=0.36     V=9352
.LBS−*.                    T=−3       N10
```

NGC 3745 is the small galaxy at 13,6. Other
galaxies in the field are: NGC 3746, the barred
galaxy at 15,5; NGC 3748, at 09,6; NGC
3750, the bright elliptical at 30,5; NGC 3753,
the edge-on evolved system at 33,5 (which
appears to have several faint star formation
regions); and NGC 3754, the barred spiral
with the bright blue ring of star formation
activity.

NGC 3756

```
797  6  79 03 26 07.1          McD 0.9m
B=12.1     B−V=0.63              V=1159
.SXT4..   Sc(s)I−II        T=4  L=3  N27
```

The small yellow nuclear region is surrounded
by a disk of relatively constant green color
with large spiral features. No blue knots are
visible.

NGC 3780

```
798  6  79 03 26 07.4          McD 0.9m
B=12.3*                         V=2865
.SAS5*.   Sc(r)II.3        T=5  L=3  N00
```

A bright nuclear region and two blue knots
stand out in this otherwise faint galaxy.

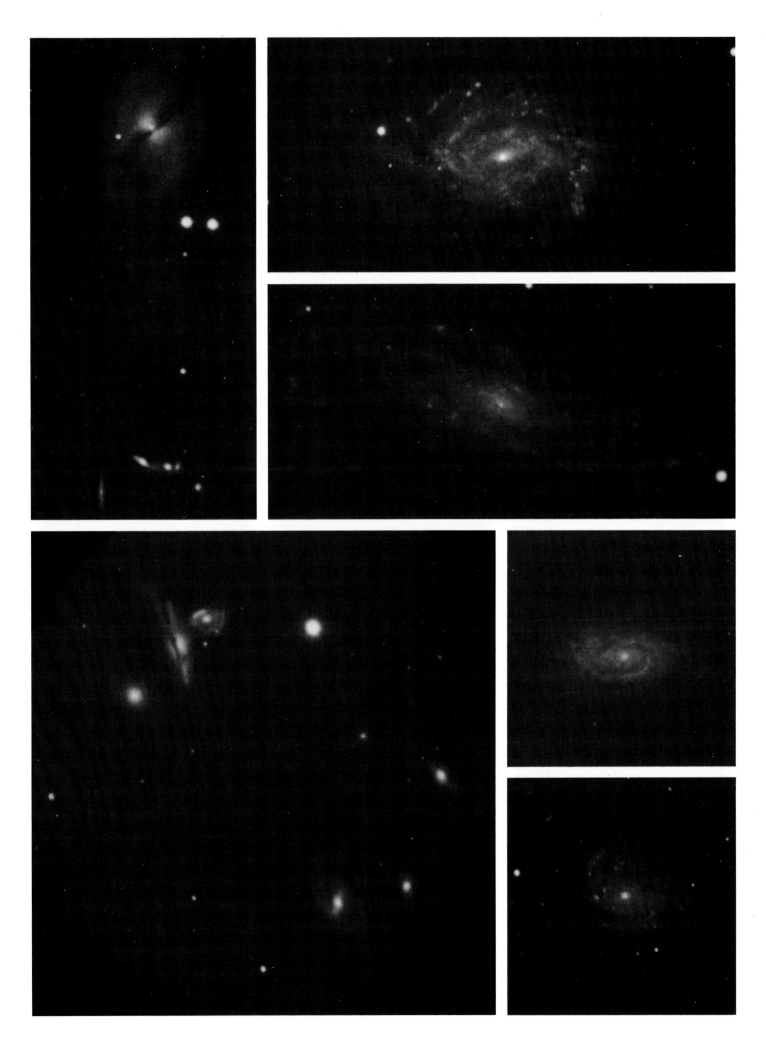

48

3786
3788
3808
3808a
3810
3893
3896
3898
3913

NGC 3786

873	3	79 04 27 05.4		McD 2.7m
				V=2745
.SXT1P.			T=1	N09

NGC 3786 on the left has a broad yellow bar in which one dust lane is visible. A partial ring of blue knots marks the edge of the inner disk. Beyond this a very faint spiral arm is visible on the original print. NGC 3788 on the right has a smaller yellow inner disk and bright spiral arms containing blue knots. Note the very bright and smooth spiral feature on the right. Refer to the discussion of NGC 4258.

NGC 3810

871	3	79 04 27 05.0		McD 2.7m
B=11.2	B−V=0.55	U−B=−.06	V=862	
.SAT5.. Sc(s)II			T=5 L=1 N27	

The nuclear region is small. Plume structure emanates directly from the bright inner disk. Note how the ends of a set of plumes comprise the loci defining the the edge of an apparent spiral arm (as in the vicinity of 15,1 to 24,2). The same kind of morphology is observed to repeat in the next arm out in the same sector (theta=15 to 24), where long plumes originating at the right tend to terminate in the line of blue knots and enhancements which comprise the long spiral arm at the bottom of the image. Note that much of this latter plume structure is traceable to the chaotic region around 07,2.

NGC 3808

831	3	79 04 25 06.1		McD 2.7m
				V=6969
.SXT5*P			T=5	N34

One spiral arm is dotted with bright blue knots whereas the other is completely lacking in them. NGC 3808a (above) has the curious color property of appearing bluer in the center than at the edge.

NGC 3893

748	6	79 03 24 06.4		McD 0.9m
B=11.1				V=1034
.SXT5*.	Sc(s)I.2		T=5 L=1* N32	

NGC 3893 is characterized by a well defined two arm spiral pattern. Blue knots appear between spiral arms at the upper left, occurring in faint plume structure. The second galaxy in the field, NGC 3896, appears to be a compact Magellanic type irregular, perhaps similar to NGC 4449.

NGC 3898

799	6	79 03 26 07.7		McD 0.9m
B=11.7	B−V=0.88	U−B=0.35	V=1137	
.SAS2..	SaI		T=2 L=3 N16	

Only the yellow old population of this galaxy is visible. No star formation regions are evident.

NGC 3913

1128	6	80 03 21 07.5		McD 2.1m
				V=952
PSAT7*.			T=7	N32

This object is similar to NGC 3227. The nuclear region is relatively bright. The low central mass argument used to attempt to explain the lack of spiral structure in NGC 2976 clearly does not hold here, yet there is no evident spiral structure. Blue knots occur in the outer regions of the disk.

A second galaxy in the field appears to exhibit a bar structure with a distinct nuclear region.

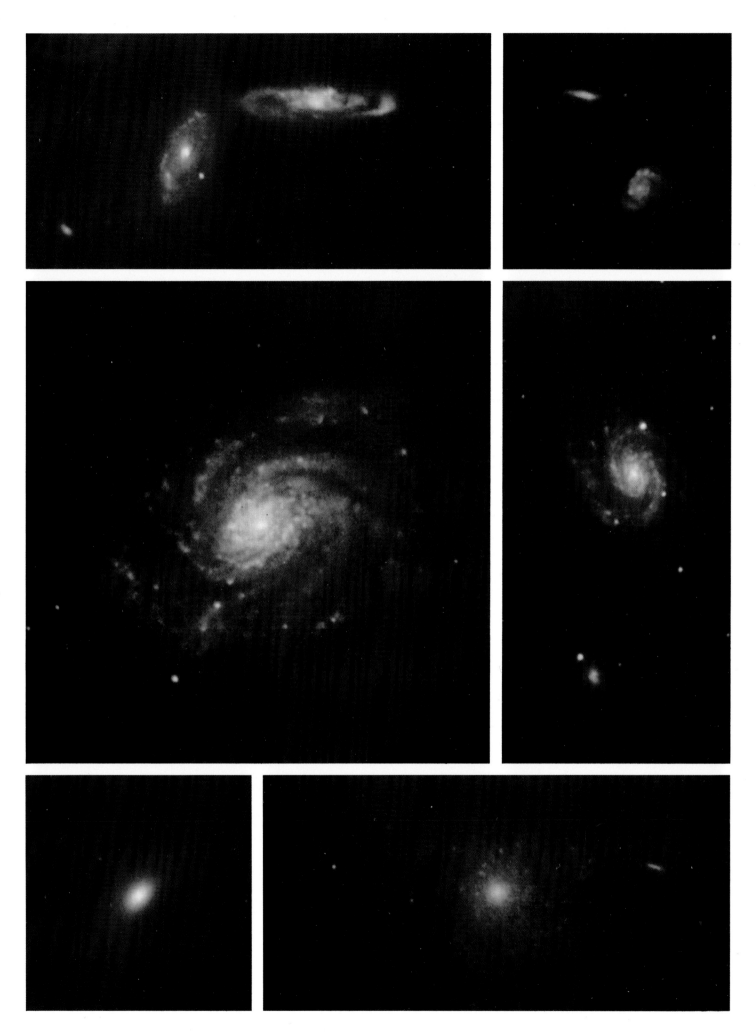

3917
3921
3930
3938
3953
3981
3992

NGC 3917

1129 6 80 03 21 07.8			McD 2.1m
B=12.4*	B−V=0.71	U−B=0.00	V=1041
.SA.6*.	Sc(s)III	T=6	N00

The disk of this galaxy is similar to that of NGC 3227, although the nuclear region is less pronounced. Evidence for spiral structure is perhaps present, but not nearly as noticeably as in NGC 3511. Blue knots are most frequent in the outer part of the disk.

NGC 3921

874 3 79 04 27 05.6			McD 2.7m
B=13.0	B−V=0.65	U−B=0.25	V=6070
PSAS0P.		T=0	N04

The smooth appearance of this remarkable highly distorted system seems to be contrary to the apparent color, yet the photoelectric value is in excellent accord with the color as illustrated. Either these stars are not of an old population or their color is greatly influenced by low metallicity effects. It seems most likely that the cataclysm that disturbed the system caused an inhibiting effect on star formation processes. Any blue knots present at that time have since evolved beyond the lifetimes of the luminous stars which gave them their color and singular high surface brightness. Thus, in this scenario, the present stellar population is of old (in the nuclear region) and intermediate age (elsewhere) only.

NGC 3930

1545 3 82 03 30 06.5		McD 2.7m
		V= 897
.SXS5..		T=5 L=6 N06

This galaxy is a very low surface brightness system containing elements of a distinct old population in the nuclear region as well as several young blue knots in the outer disk.

NGC 3953

489 6 78 01 09 09.8			McD 0.7m
B=10.7	B−V=0.70	U−B=0.10	V=1043
.SBR4..	SBbc(r)I−II		T=4 L=1 N26

Note the bar-end enhancements. See discussion of NGC 151 and NGC 936. The system has numerous spiral arms which broaden with increasing distance from the center. These arms are well defined but mostly of intermediate color. Blue knots are present throughout most of the disk.
Other faint galaxies in the field appear to belong to a distant background cluster.
A number of these galaxies are quite blue.

NGC 3938

752 6 79 03 24 07.7			McD 0.9m
B=10.9	B−V=0.52	U−B=−.10	V=838
.SAS5..	Sc(s)I		T=5 L=1 N07

The yellow nuclear region is surrounded by moderately well defined spiral arms, comprised largely of a series of straight segments. Blue knots extend beyond the visible spiral structure. Faint plume structure is evident.

NGC 3981

1269 6 81 05 07 02.1		LC 1.0m
B=12.4*		V=1561
.SAT4..	Sb:(I)	T=4 N25

The long spiral arms and their relationship to the bar are similar to those of NGC 6872.

NGC 3992   M109

749 6 79 03 24 06.8		McD 0.9m
B=10.6	B−V=0.79	V=1149
.SBT4..	SBb(rs)I	T=4 L=1 N32

The left bar-end enhancement is double, probably cut by a dust lane. Faint blue ring and arm structures cross the bar axis just outside the bar-end enhancements. Except for its much lower surface brightness this galaxy is similar in many respects to NGC 151.

4026
4030
4038
4051
4085
4088
4094

NGC 4026

1130	6	80 03 21 08.1		McD 2.1m
B=11.7				V=958
.L..../	S0₁/₂(9)		T=−2	N09

The tapering of the ends to near points shows that this is an edge-on view of a system with a disk and not a face-on view of a bar. Thus the apparent lack of dust implies that the system is in fact essentially devoid of dust. There is no evidence of recent star formation. This is the type example of an S0 galaxy (disk, no evident dust and no recent star formation). See also NGC 3115.

NGC 4030

1273	6	81 05 07 03.2		LC 1.0m
B=11.1*				V=1255
.SA54..	Sbc(r)I		T=4 L=1	N16

A bright intermediate population dominates the central region of the galaxy, making the overexposed nuclear region difficult to assess. Star formation is occurring in many of the spiral features.

NGC 4051

490	6	78 01 09 10.2			McD 0.7m
B=10.9	B−V=0.67	U−B=0.00		V=726	
.SXT4..	Sbc(s)II			T=4 L=3	N04

This is one of the classical Seyfert galaxies. The star-like nucleus is not evident here due to the brightness of the overall nuclear region. The central region contains a broad loosely developed bar of old stars. The arms are bright and well developed with both intermediate and young stars. See below.

NGC 4038

1293	6	81 05 08 02.7			LC 1.0m
B=11.3	B−V=0.60	U−B=−.22		V=1447	
.SBS9P.	Scp			T=9	N25

The nuclear regions of these two galaxies are easily identifiable by their color. Note that one spiral arm of the galaxy in the center is folded back on the other at 18,1. One spiral arm from the other galaxy (NGC 4039) also passes through the same region. Note the displacement of the spiral patterns from the smooth yellow disk populations of these galaxies. A truly cataclysmic collision.

NGC 4051

1131	1.5	80 03 21 08.3			McD 2.1m
B=10.9	B−V=0.67	U−B=0.00		V=726	
.SXT4..	Sbc(s)II			T=4 L=3	N04

The Seyfert nucleus is still bright and overexposed in this short exposure. In the spiral arms the blue knots are singled out. Note the small size of most of these individual regions of star formation.

NGC 4085

875	3	79 04 27 05.7		McD 2.7m
B=12.9	B−V=0.59			V=764
.SXS5*	$ ScIII:		T=5 L=5*	N04

The nuclear region is small. Star formation dominates this galaxy.

NGC 4088

275	6	76 04 01 09.8		McD 0.9m
B=11.1	B−V=0.60			V=822
.SXT4..	Sc(s)II−III/SBc		T=4 L=2*	N32

The spiral arm below and to the left originates at the right end of the yellow bar, while the upper arm originates at the left. In a sense the structure is analogous to that of NGC 1187, but in mirror image. Star formation extends well into the bar region.

NGC 4094

1272	6	81 05 07 02.9		LC 1.0m
B=12.5*				V=1229
.SAT5*.	Sbc(s)II		T=5 L=7	N00

The nuclear region of this galaxy is the site of some activity with diffuse red, yellow and blue knots in close proximity. The disk lacks definitive spiral structure; a bright blue knot is visible near its edge.

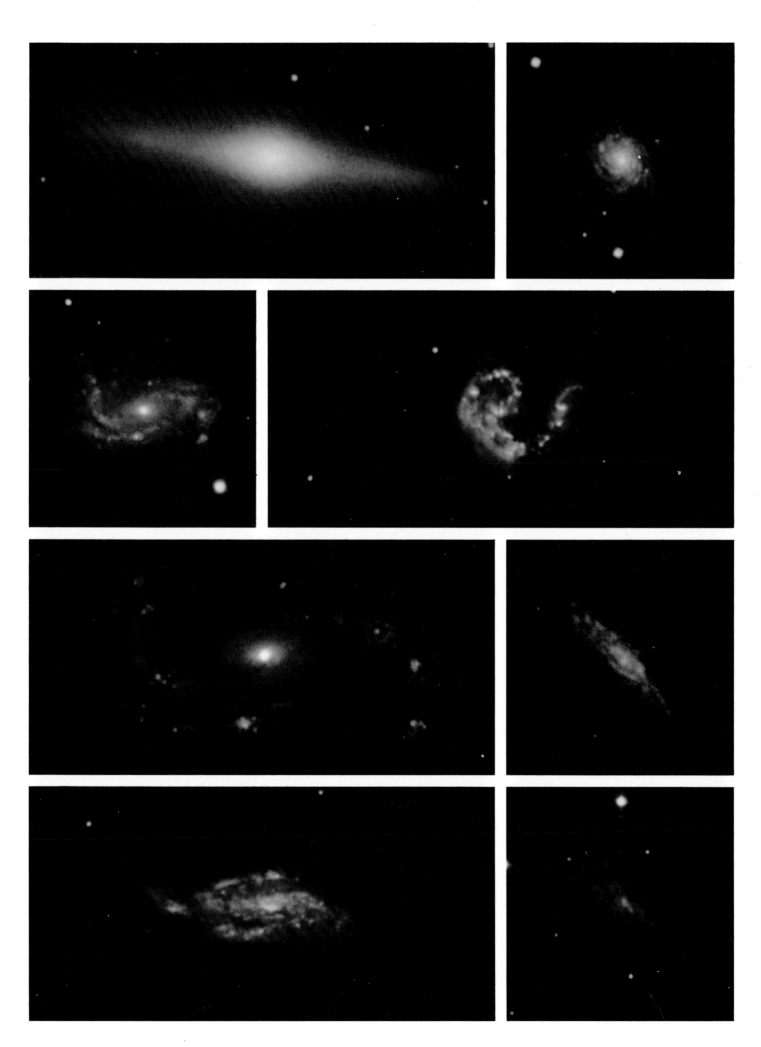

4096
4100
4123
4136
4144
4145
4151

NGC 4096

911b 6 79 04			McD 0.9m
B=11.0	B−V=0.44	U−B=0.20	V= 561
.SXT5..	Sc(s)II−III		T=5 L=3* N09

Dust lanes and spiral features dominate the disk of this system. Blue knots are visible throughout the disk.

NGC 4100

1546 3 82 03 30 06.5			McD 2.7m
B=11.6	B−V=0.73	U−B=0.06	V=1156
PSAT4..	Sc(s)I−II		T=4 L=2* N27

Some structure is apparent in the nuclear region. The galaxy appears excessively red (a 1500 series photograph; see NGC3705), but faint star formation regions (blue) are visible in much of the spiral structure, and two regions are prominent in the blue light of young stars.

NGC 4144

913b 6 79 04			McD 0.9m
B=11.7	B−V=0.43	U−B=−.20	V=324
.SXS6$/	ScdIII		T=6 L=5 N00

A considerable intermediate age population is in evidence here, together with numerous faint blue knots. There is an indication of a small, faint yellow nucleus.

NGC 4145

220 6 76 03 31 08.2			McD 0.9m
B=11.5	B−V=0.54	U−B=−.10	V=1035
.SXT7..	SBc(r)II		T=7 L=3 N00

The short yellow bar appears to lack a central nucleus. Faint spiral arms are delineated in some regions by numerous blue knots. In other areas an occasional bright blue knot is visible. Compare with NGC 5669.

NGC 4123

1521 6 82 03 28 04.6			McD 0.7m
B=11.8	B−V=0.62	U−B=−.07	V=1189
.SBR5..	SBbc(rs)		T=5 N00

The nucleus stands out in sharp contrast to the faintness of this galaxy. A faint yellow bar and broad blue ring are discernible.

NGC 4136

753 6 79 03 24 08.0			McD 0.9m
B=11.6*			V=434
.SXR5..	Sc(r)I−II		T=5 L=5 N00

The bright nuclear region appears to be attended by a faint bar with yellow enhancements in the inner ring. Blue knots are visible in the outlying structure.

NGC 4151

914b 6 79 04			McD 0.9m
B=11.1	B−V=0.75	U−B=0.00	V=1002
PSXT2*.	Sab		T=2 N00

This galaxy has a bright, dominant yellow nuclear region, together with a curious bar-like appearance, with the bar lobes dominated by a young blue population. The bar is not bright but its size is on the order of galactic dimensions. This feature is perhaps relevant in regard to the two very peculiar galaxies NGC 3561a and NGC 6054.

NGC 4151

1134 1.5 80 03 21 08.8			McD 2.1m
B=11.1	B−V=0.75	U−B=0.00	V=1002
PSXT2*.	Sab		T=2 N00

This is a Seyfert galaxy; and this short exposure taken with a larger telescope than was used for the above photograph was intended to reveal more of the galaxy's bright nuclear region. Two dust lanes are visible, but the central region is still too overexposed to show either the color or the star-like appearance of the brilliant active nucleus of this galaxy.

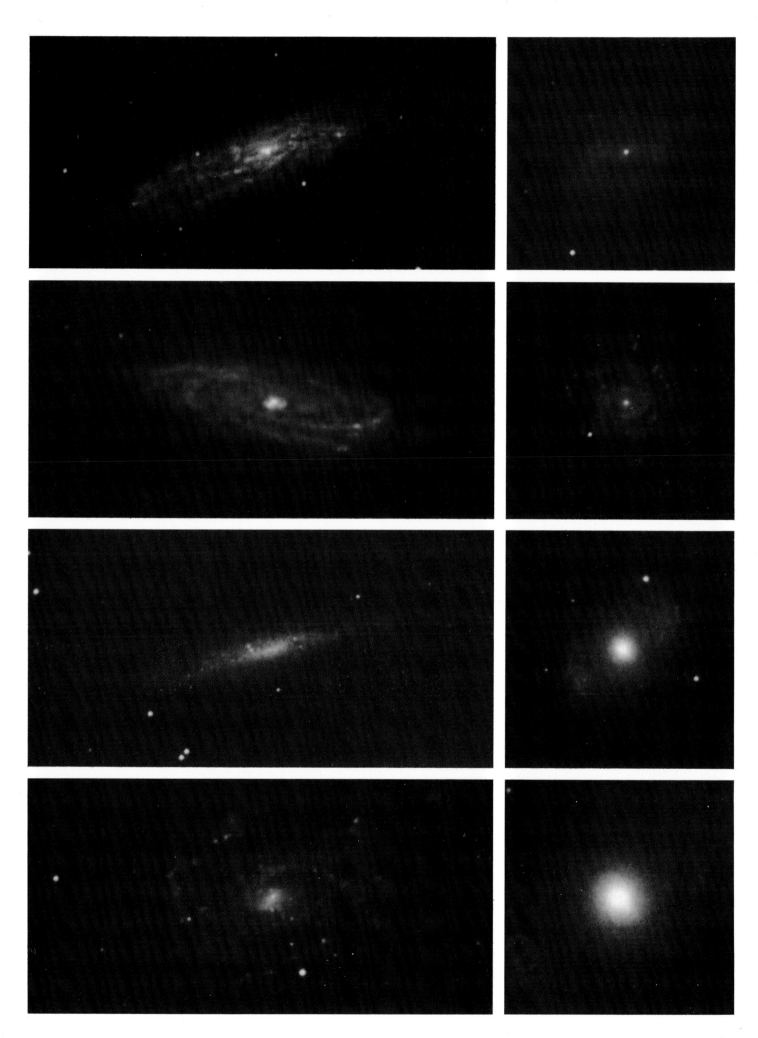

4157
4162
4178
4183
4189
4192

NGC 4157

915b	6	79	04		McD 0.9m
B=11.9	B−V=0.75		U−B=0.10		V=907
.SXS3$/	Sbc		T=3	L=3*	N14

This photograph and the one to the upper right provide another comparison of results obtained under different circumstances. In this case two different telescopes and two different image tubes were used.

Corwin's photometry of the bright star is V=8.06, B−V=0.49 and U−B=0.02.

NGC 4157

1547	3	82 03 30 06.8			McD 2.7m
B=11.9	B−V=0.75		U−B=0.10		V=907
.SXS3$/	Sbc		T=3	L=3*	N14

The yellow nuclear region and inner disk are evident and appear slightly reddened due to dust. A region of star formation (apparently also reddened) is visible to the left. The outer arms are blue with young stars. This photograph is slightly bluer in color balance than the norm for the 1500 series. (See NGC 3705).

NGC 4162

832	3	79 04 25 06.3			McD 2.7m
B=12.2	B−V=0.80		U−B=0.30		V=2454
RSAT4..	Sc(s)I−II		T=4	L=3	N09

This galaxy is rather similar to NGC 3344 and NGC 3810.

NGC 4183

916b	6	79	04		McD 0.9m
B=12.4*					V=981
.SAS6*/	Scd		T=6	L=7*	N04

The appearance of this nearly edge-on system is somewhat similar to that of NGC 4244 although not quite as blue. A weak dust lane is visible. Note the blue knot at 0.6,3

NGC 4178

754	6	79 03 24 08.3			McD 0.9m
B=11.8	B−V=.50		U−B=−.10		V=202
.SBT8..	SBc(s)II		T=8	L=3*	N28

This photograph and the one at the right comprise another control pair. The narrow bar appears rather straight but discontinuous. Faint spiral arms are visible.

NGC 4178

1580	3	82 03 31 07.7			McD 2.7m
B=11.8	B−V=.50		U−B=−.10		V=202
.SBT8..	SBc(s)II		T=8	L=3*	N28

Comparison with the photograph at the left reveals the characteristic color bias associated with the 1500 series photographs. The interpretation in terms of intermediate or old population by means of green versus yellow is clearly biased towards the yellow here. It is important to assess these variances for yourself using photoelectric data provided throughout the atlas, both for stars and galaxies, that you be able to judge your own interpretations of these photographs as well as those presented in the atlas text.

Note that at this larger scale it is possible to identify star formation regions within the bar, and the three large outlying blue knots are further resolved into smaller blue knots.

NGC 4189

1584	3 82 03 31 08.6			McD 2.7m
B=12.5 B−V=0.77				V=2013
.SXT6$.	SBc(sr)II.2	T=6	L=3*	N00

The bright yellow nucleus appears to have a short faint bar. Several blue knots are visible in the outer spiral structure.

NGC 4192   M 98

918b	6	79	04		McD 0.9m
B=10.8	B−V=0.79		U−B=0.18		V=−206
.SXS2..	SbII:		T=2	L=2*	N24

The nuclear region has a tendency towards the box appearance. Note the extreme reddening due to dust in the entire darkened area of the disk especially to the left of the nucleus. Blue knots are in evidence, widely scattered as is frequently the case.

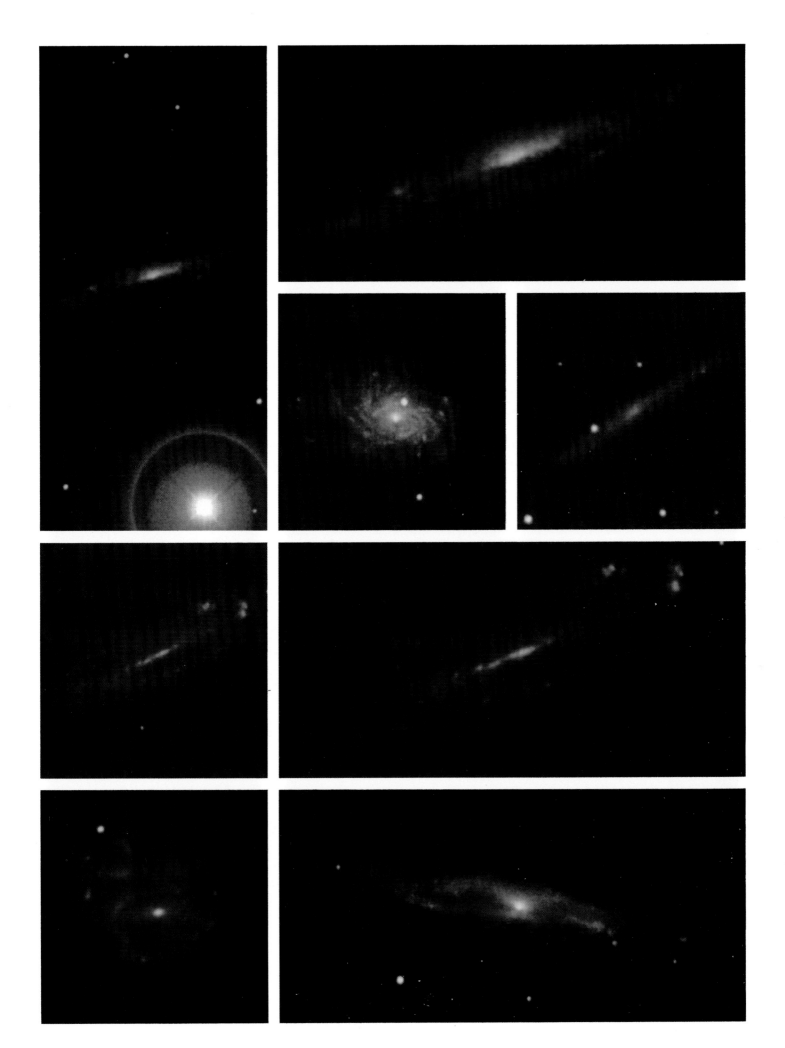

NGC 4214

72 10 75 05 03 05.6			McD 0.9m
B=10.2	B−V=0.46	U−B=−.30	V=309
.IXS9..	SBmIII		T=0 L=6 N00

A smooth yellow-green population underlies a massive burst of star formation activity extending all the way into the central region. No nucleus is visible. Supergiants may be at the threshold of detection. The greenish smooth population probably contains generally old stars, but the color does not prove their existence as would the yellow color seen in the other three galaxies illustrated on this page. See also NGC 4449.

NGC 4216

919b 6 79 04			McD 0.9m
B=10.9	B−V=0.99	U−B=0.55	V=−69
.SXS3*.	Sb(s)		T=3 L=3 N00

This galaxy has the peculiar property of a yellow brightening near the edge of the yellow disk (34,2). This feature could be a bar-end enhancement, not matched on the other end because of intervening dust. The fact that similar phenomena are seen in nearly edge-on systems (e.g. NGC 5170) indicates that perspective is a factor, and obscuring effects due to dust could be considered a possible, although not entirely compelling, explanation. The faint outer disk contains massive amounts of dust.

NGC 4217

1136 6 80 03 21 09.0			McD 2.1m
B=11.3*	B−V=0.85	U−B=0.30	V=1054
.S..3./	Sb:		T=3 N28

Notice the inclined chains of blue knots embedded in the dust lane, as if the entire disk was gently undulating. The old yellow population is dominant above and below the plane over much of the disk. This effect would appear less significant from a more face-on perspective.

Corwin's photometry of the bright yellow star is V=9.02, B−V=0.94 and U−B=0.64.

NGC 4219

1234 6 81 05 04 02.5			CTIO 0.9m
B=12.3	B−V=0.80	U−B=0.10	V=1694
.SAS4..	Sbc(s)(II−III)		T=4 N00

The disk of this galaxy is quite red, and is crossed with dust lanes in several directions.

54

4236
4242
4244
4254
4258
4274
4293

NGC 4236

491 6 78 01 09 10.5			McD 0.7m
B=9.9	B−V=0.40	U−B=−.15	V=160
.SBS8..	SBdIV		T=8 L=7 N26

The bar center is distinctly yellower than the ends, indicating an evolutionary progression in the stellar population. Hence there probably are old stars in this system, but their contribution is dwarfed by the young stars, blue knots and brighter stars of intermediate age. This galaxy is resolved into supergiants at a distance modulus by visual inspection of approximately +27. See NGC 625.

NGC 4242

775 6 .79 03 25 06.8			McD 0.9m
B=11.6	B−V=0.55	U−B=−.13	V=749
.SXS8..	SBdIII		T=8 L=7 N07

The bright blue knot in this galaxy outshines its greenish appearing nucleus. Other faint blue knots are scattered throughout the field (e.g. 24,3).

NGC 4244

921b 6 79 04		McD 0.9m
B=10.5	B−V=0.44	V=270
.SAS6*/	Scd	T=6 L=7* N12

This system is remarkable not only for its profusion of intermediate aged stars, but also for the considerable amount of star formation activity with very little dust. Unlike many systems this one does not have an external band of dust.

NGC 4254   M 99

471 6 78 01 08 11.6			McD 0.7m
B=10.4	B−V=0.58	U−B=−.02	V=2324
.SAS5..	Sc(s)I.3		T=5 L=1 N09

The yellow nucleus is overexposed and nearly lost in the glare of light from bright inner star formation regions and intermediate aged stars. Massive bursts of star formation are occurring in the major spiral arm regions and in plume structures (above) as well.

NGC 4258   M 106

75 10 75 05 03 06.7			McD 0.9m
B=8.9	B−V=0.68		V=537
.SXS4..	Sb(s)II	T=4	N06

The yellow nuclear region is surrounded by a smooth disk, yellow in some places and green in others. Narrow strings of blue knots delineate spiral arms which begin near the perimeter of the disk.

A bright smooth green region at 08,2 is a chief point of interest. Its color is clearly intermediate between the nuclear population and the blue knots. Since a yellow population is present we know that subsequent populations will not be metal-poor, hence this green feature is definitely of intermediate age. It lies in a spiral feature, but is considerably removed from the zone of star formation (the blue knots and the lane of dust just interior to them). This object is a remnant spiral feature, a fossil spiral arm, gradually beginning to disperse some hundreds of millions of years after it was formed in a massive burst of star formation activity. There are numerous other examples of fossil spiral features in the atlas, but this one serves as the definitive prototype.

NGC 4274

776 6 79 03 25 07.0			McD 0.9m
B=11.2	B−V=0.93	U−B=0.41	V=715
RSBR2..	Sa(s)		T=2 L=4 N16

Dust lanes reveal the existence of a central curving bar, interior to a broad ring structure. Another galaxy is visible at 31,4.

NGC 4293

1548 3 82 03 30 06.8			McD 2.7m
B=11.1	B−V=0.92	U−B=0.35	V=825
RSBS0..	Sap	T=0	N00

The disk population is smooth and mostly yellow. Complex swirls of dust appear to have a global pattern, possibly with significant out-of-plane components. A blue knot is visible near the end of one particularly dense dust lane.

The diffuse red object at 29,3 is very probably a background galaxy seen through the disk of NGC 4293.

NGC 4298

492 6 78 01 09 11.0   McD 0.7m
B=12.0   B−V−0.71   V=1042
.SAT5..   Sc(s)III   T=5   N27

NGC 4298 (above) is intermediate in appearance to NGC 2976 and NGC 3227. There is little if any spiral trend to the patchy pattern in the disk, but star formation activity is widespread nevertheless. Clearly this star formation activity must be caused by stochastic processes. But why is the activity of these processes distributed so uniformly on the galactic scale? Also it appears that in the past star formation either occurred more frequently in the central region to account for its predominantly older population, or (improbably) a relaxation mechanism worked to draw in and trap stars in the inner region over time scales on the order of billions of years.

We have a perfect edge-on view of NGC 4302, in which we see that the yellow nuclear region does not bulge. Instead the disk thickness remains nearly constant while transitioning through color and population types ranging from yellow and old near the center to blue and young in the outer regions. Note the blue knot at the right.

NGC 4303   M 61

1139 3 80 03 21 09.7   McD 2.1m
B=10.2   B−V=0.54   U−B=−.11   V=1483
.SXT4..   Sc(s)I.2   T=4 L=1 N08

This magnificent barred spiral is undergoing an incredible burst of star formation activity. The bar ends curve into the arm structure. Note how the upper arm tends to flow from the right bar end, and the lower from the left end, but both arms also extend inward past the bar ends to engage the entire leading edge portions of the bar. The bright arm segment at the right is an extension of the lower arm. A broad but poorly defined dust lane separates plume structure along the entire lower edge of the bar from the arm segment below and to the right. Using the dust lanes as arm separators it is clear that this plume structure along the lower bar edge is actually related to the upper arm, as is the upper bar edge to the lower arm.

NGC 4314

1138 1.5 80 03 21 09.4   McD 2.1m
B=11.3   B−V=0.84   U−B=0.30   V=879
.SBT1..   SBa(rs)p   T=1   N35

In this photograph only a bar is seen, but in the center of the bar there is a bright spiral pattern. Dust lanes are associated with the spiral, and give rise in part to its appearance. Bright knots are present, but are too overexposed to appear blue.

NGC 4314

1138 1.5 80 03 21 09.4   McD 2.1m
B=11.3   B−V=0.84   U−B=0.30   V=879
.SBT1..   SBa(rs)p   T=1   N35

This print was made using a subtractive color masking technique to artificially enhance the color. The result reveals the presence of a ring of star formation regions surrounding the nucleus, and a spiral pattern which is different in color from its outer surroundings. The dust lanes add slight reddening but do not account entirely for this difference in color. Hence the spiral is interpreted as an intrinsic stellar formation with a slightly different stellar population composition than in the main body of the bar. See also discussions of NGC 613, NGC 1365, and others referred to therein. The orientation of this enlarged scale print is a (r/l) mirror image of the normal image (above). This galaxy could be considered a type example for an evolved system with second wave star formation activity. Refer to the discussion of NGC 4826.

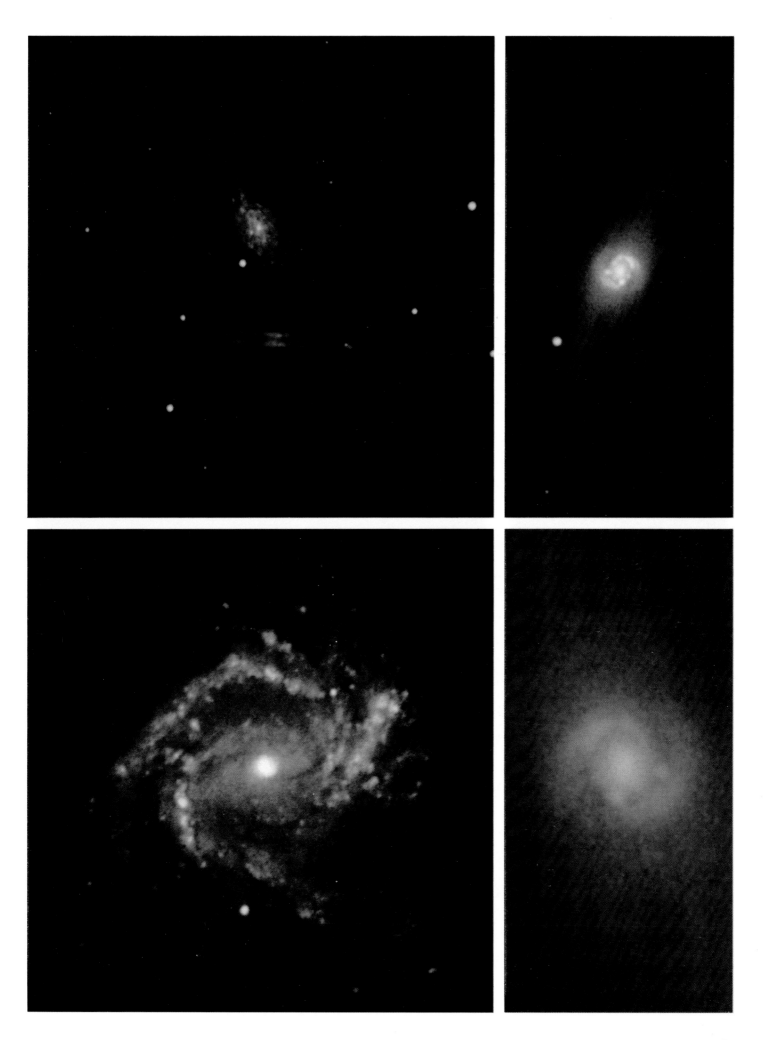

NGC 4321 M 100

472 6 78 01 08 12.0			McD 0.7m
B=10.1	B−V−0.73	U−B=−.06	V=1543
.SXS4..	Sc(s)I		T=4 L=1 N27

This beautiful spiral galaxy has a very bright nuclear region surrounded by a relatively faint and yellow inner disk. Blue spiral arms originate outside this disk and are densely populated with bright blue knots. One of the brightest blue knots occurs at the terminus of a string of star formation regions in the arm at the right, yet the faint intermediate population extends well beyond this point in the arm.

NGC 4365

1553 3 82 03 30 07.6			McD 2.7m
B=10.6	B−V=0.96	U−B=0.52	V=1074
.E.3...	E3		T=−5 N04

Note the contrast in overall properties between this galaxy and NGC 4321. NGC 4365 appears to be comprised of only the old yellow stellar population. No dust is evident in this photograph, but its presence in small amounts is not ruled out.

NGC 4321   M 100

835 3 79 04 25 07.8			McD 2.7m
B=10.1	B−V=0.73	U−B=−.06	V=1543
.SXS4..	Sc(s)I		T=4 L=1 N27

This photograph at higher resolution was taken at a time near maximum brightness of the supernova which occurred in NGC 4321 in 1979, the bright blue star at 12,6.
In the galaxy, note among other things the pattern of dust lanes in the lens. This pattern is actually characteristic of dust lanes associated with bars. The tendency for spiral arms to form outside a bar region may also be relevant here. The implication is that this galaxy contains an insipient barred structure. It is conceivable that that this system may actually be in dynamical transition into a fully developed barred galaxy.

NGC 4321   M 100

1084 0.5 80 03 18 10.8			McD 2.1m
B=10.1	B−V=0.73	U−B=−.06	V=1543
.SXS4..	Sc(s)I		T=4 L=1 N27

This very short exposure at nearly the same scale as the one on the left shows a ring of active star formation regions deep in the nuclear region. In the center a small yellow nucleus is visible. The nuclear pattern appears to be influenced by factors relating to the dust lanes as well as by the star formation regions themselves. Hence, because patterns of dust lanes appear to be globally associated in general, the implication arises that energy mechanisms in nuclear regions may influence overall galactic structure.

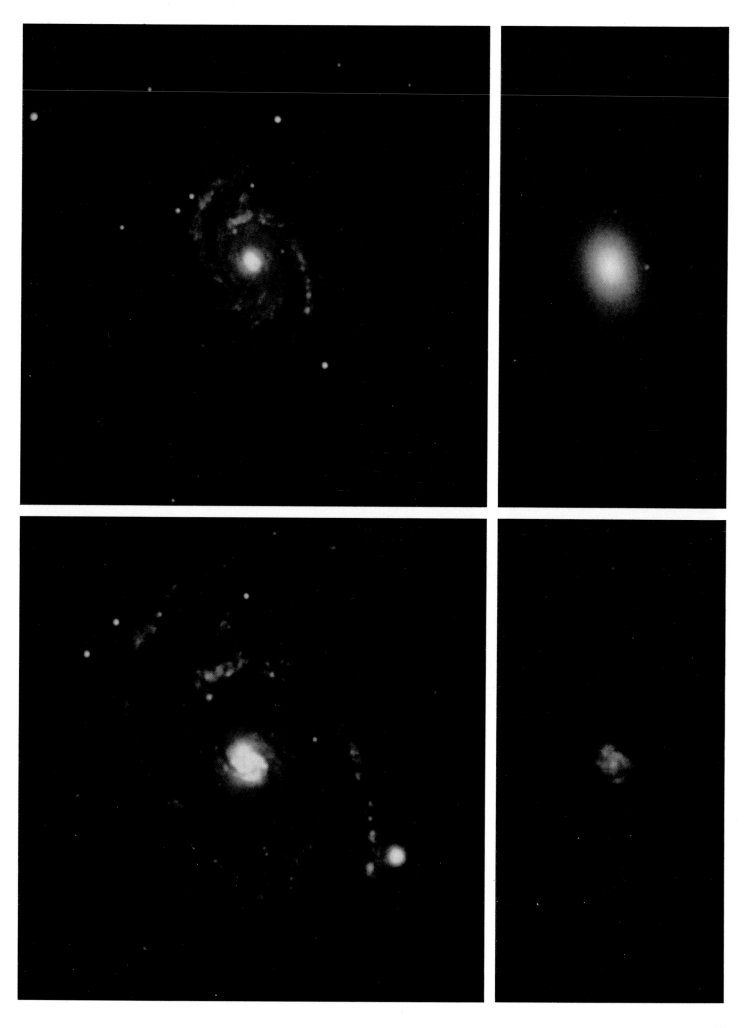

NGC 4374 M 84

1091	6	80	03	19	08.0			McD 2.1m
B=10.3		B−V=0.97		U−B=0.58			V=854
.E.1...		SB02/3(r)(3)			T=−5			N00

This photograph of a giant elliptical galaxy gives no indication of the presence of anything other than the old yellow stellar population. The red object at 28.2 is a plate defect.

NGC 4382 M 85

1549	3	82	03	30	07.1			McD 2.7m
B=10.1		B−V=0.88		U−B=0.40			V=718
.LAS+P.		S01(3)p			T=−1			N00

This system also appears to consist of old population stars only.
The dark patch at 35,2 is a defect present on plates numbered 1540 and greater.

NGC 4394

1550	3	82	03	30	07.2			McD 2.7m
B=11.7		B−V=0.82		U−B=0.25			V=717
RSBR3..		SBb(sr)I−II			T=3	L=3	N35

The old yellow bar dominates this system. A faint patchy ring is visible.

NGC 4395

922b	6	79			05.9			McD 0.9m
B=10.6		B−V=0.54		U−B=−.25			V= 307
.SAS9*.		SdIII−IV			T=9	L=8	N00

The center of spiral symmetry of this galaxy is near the white star at 07,1. The faint object just to the left of that star may be the nucleus, if there is one at all. The visible galaxy is comprised almost entirely of star formation regions – blue knots. Most of the star formation regions are a considerable distance from the center. This pattern is clearly inefficient for generating a large central population of stars. What constrains this system to such a slow rate of star formation? Is this perhaps a relatively recent addition to the galaxian census? Knowledge of the nature of the nucleus could provide the answer to these questions.

4402
4406
4414
4429
4435
4438

NGC 4402

214	6	76	03	31	07.0		McD 0.9m
B=12.5		B−V=0.83		U−B=0.06		V=73	
.S..3./						T=3	N00

The peculiar color asymmetry appears to be due to an inner dust ring. The outer disk is smooth and somewhat green, similar perhaps to NGC 4244. Several blue knots are evident. The bright white object near the left edge of the galaxy is a supernova.

NGC 4406   M 86

1552	3	82	03	30	07.5		McD 2.7m
B=10.1		B−V=0.93		U−B=0.52		V=−419	
.E.3...		S01(3)/E3				T=−5	N00

There is no sign of young or intermediate aged stars or dust in this galaxy. The dark spot at 34,2 is an image defect.

NGC 4414

1582	3	82	03	31	08.1		McD 2.7m
B=10.9		B−V=0.69		U−B=0.1		V=718	
.SAT5$.		Sc(sr)II.2				T=5	L=3* N23

The bright inner disk exhibits considerable structure, but is dominated by the light of yellow stars. The dark lanes appear reddened suggesting that a significant amount of this structure is due to dust. Plume structure dominates the morphology of the outer disk.

NGC 4429

778	6	79	03	25	07.6		McD 0.9m
B=11.1		B−V=0.94		U−B=0.54		V=1029	
.LAR+..		S03(6)Sap				T=−1	N04

There appears to be a small dust ring in the nuclear region. Two enhancements at the lower right and upper left have the appearance of bar-end libration nodes. See NGC 1079 and the discussion associated with NGC 936. Corwin's photometry of the bright star gives V=9.07, B−V=0.56 and U−B=0.09

NGC 4435

809	6	79	03	26	10.6		McD 0.9m
B=11.7		B−V=0.92		U−B=0.48		V=793	
.LBS0..		SB01(7)				T=−2	N09

NGC 4435 (rignt) appears to consist of old yellow stars. Several small defects give the appearance of dust. The pattern of dust in NGC 4438 is real, however, as shown in the larger scale photograph below.

NGC 4438

1583	3	82	03	31	08.4		McD 2.7m
B=10.8		B−V=0.83		U−B=0.37		V=182	
.SAS0P*		Sb(tides)				T=0	N09

The pattern of dust lanes crossing this galaxy is not entirely unique. In fact there is a small sub-class of galaxies of similar appearance (see especially NGC 4753, and also NGC 2146). These are tidally disturbed systems. Ordinary dust lanes in the disk become intricate quilt-work patterns when distorted and bent out of plane by tidal forces and viewed in projection against the bright galactic background.

756   6   79 03 24 08.8   McD 0.9m
B= 9.8   B−V=0.41   U−B=−.30   V=262
.IB.9..   SmIV   T=10 L=5 N00

If an old yellow population is present, it is well
masked in the smooth green population seen
here. Hence, it is not altogether inconceivable
that this system does not contain any old stars;
i.e. that it is intrinsically younger, not simply
less evolved, than systems with yellow stellar
populations. Alternatively, the old stars may
be metal-poor and appear blue-green as a
result. This hypothesis is based on at least one
blue appearing dwarf elliptical galaxy which is
a companion of a galaxy with a yellow
population component (e.g. the companion to
NGC 4631). This is a problem which needs to
be resolved.

There is a massive burst of star formation
activity going on at present in this galaxy.
Close examination of the faintest images
shows that individual supergiant stars of
various colors appear to be resolved. On this
basis we may make an approximate informal
estimate of the distance modulus, as we have
for other galaxies throughout the atlas (see
NGC 604, NGC 625 and others). In this case
the uncorrected blue-magnitude distance
modulus would be about +27.

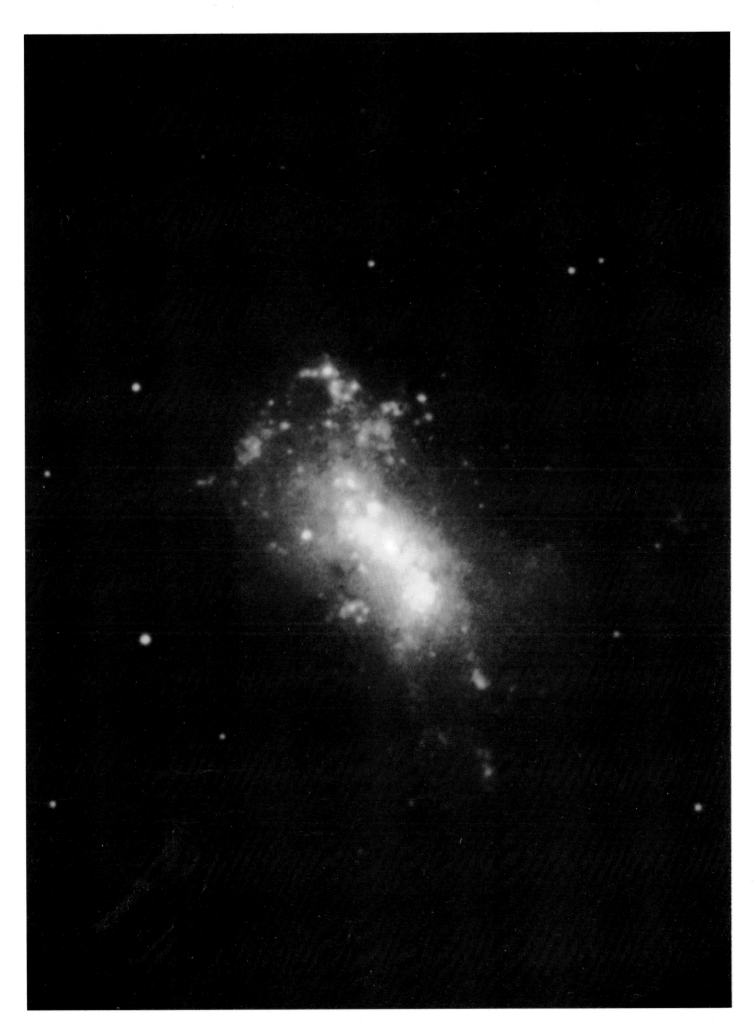

4450
4472
4486
4487
4490
4485
4496
4501

NGC 4450

1551	3	82 03 30 07.3			McD 2.7m
B=10.9	B−V=0.82				V=1990
.SAS2..	Sabp		T=2		N27

This galaxy is dominated by the old yellow population, but a bright blue knot indicates the presence of star formation activity. The pattern of dust lanes is interesting. Two lanes diverge at nearly a right angle from an apparent point of bifurcation at 30,2.

NGC 4472   M 49

1529	6	82 03 28 07.4			McD 0.7m
B= 9.3	B−V=0.98	U−B=0.56			V=817
.E.2...	E1/S0₁(1)		T=−5		N09

A system comprised entirely of the old yellow stellar population.

NGC 4486 M 87

1086	0.5	80 03 18 11.0			McD 2.1m
B=9.56	B−V=0.94	U−B=0.57			V=1409
CE.0...	E0		T=−6		N00

This is a short exposure of the giant elliptical galaxy, M 87, showing the 'jet' which appears blue. Interest in this galaxy was heightened by suggestions that its known properties are consistent with the presence of a 'black hole' at its center.

NGC 4487

1274	6	81 05 07 03.6		LC 1.0m
B=11.7*				V=876
.SXT6..	SBc(s)II.2		T=6 L=4 N09

This galaxy has a widespread smooth green population which defines the visible spiral pattern. Blue knots are prominent, but not in sufficient numbers to delineate the spiral structure. The red object is a plate defect.

NGC 4490

757	6	79 03 24 09.1		McD 0.9m
B=10.2	B−V=0.44	U−B=−.18		V=629
.SBS7P.	ScdIIIp		T=7 L=5 N15

The central region appears to be tinted with yellow, suggestive of the presence of an old yellow population mostly masked by the bright young blue and intermediate aged green stellar populations which dominate the system. The 'disk' is similar in color to that of NGC 4449, but exhibits the patchy quality seen in NGC 2976. The smaller galaxy (NGC 4485) appears to be essentially identical in population composition to the larger galaxy, and the combined system may represent a prior single galaxy which has been partially disrupted.

NGC 4496

1554	3	82 03 30 07.7		McD 2.7m
B=11.9	B−V=0.50	U−B=−.15	V=1651
SBT9..	SBcIII−IV		T=9 L=5* N33

Very bright blue knots occur in the field of these two galaxies. Both galaxies have yellow central regions which also support star formation activity.

NGC 4501   M 88

493	6	78 01 09 11.4		McD 0.7m
B=10.2	B−V=0.75	U−B=0.25	V=1989
.SAT3..	Sbc(s)II		T=3 L=1 N22

Numerous narrow arms are sharply defined by small blue knots. The outer arms at the top are excellent examples of fossil-arm structures. See NGC 4258.

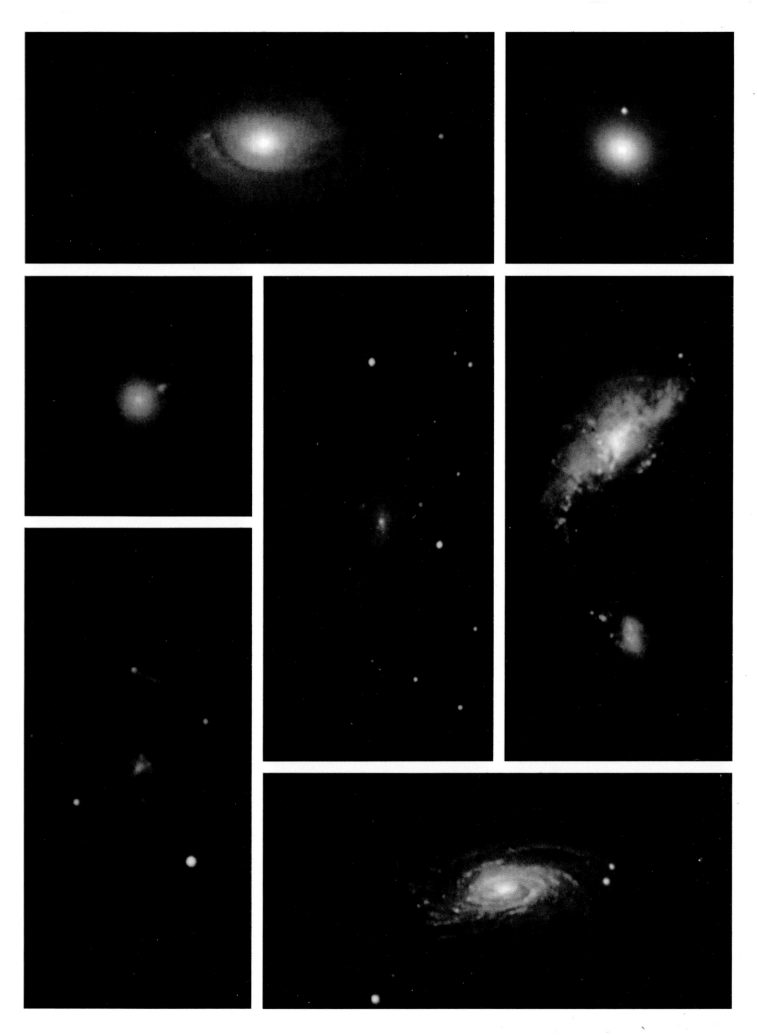

4504
4517
4526
4527
4535
4548

NGC 4504

1556	3	82 03 30 08.3		McD 2.7m
B=11.9*				V=839
.SAS6..	Sc(s)II		T=6	N12

A very low surface brightness galaxy with a distinct old nuclear region as well as faint blue knots. A blue stellar object is seen at 20,4.

NGC 4517

1531	6	82 03 28 08.0		McD 0.7m
B=12.6	B−V=0.48			V=1001
.SBT8*.	Sc		T=8 L=9	N18

Similar to NGC 4504, this large faint galaxy with an old central population continues to support star formation as evidenced by the two conspicuous blue knots.

NGC 4526

755	6	79 03 24 08.5		McD 0.9m
B=10.5	B−V=0.94	U−B=0.54		V=355
.LXS0*.	S03(6)		T=−2	N21

This galaxy appears to be comprised of the old population only. The shape may be related to the 'box' structure found in some galaxies. Note, however, the skew in the orientation of the nuclear region in this galaxy and the associated asymmetry of the disk. These features seem to be consistent with a barred spiral with the bar seen nearly end-on. See NGC 1023.

NGC 4527

1555	3	82 03 30 07.9		McD 2.7m
B=11.3	B−V=0.88	U−B=0.20		V=1614
.SXS4..	Sb(s)II		T=4 L=3	N17

In addition to the old yellow population this system contains a great deal of dust causing internal reddening and extinction. A blue knot is visible at 10,3.

NGC 4548   M 91

807	6	79 03 26 10.0		McD 0.9m
B=10.9	B−V=0.79	U−B=0.30		V=403
.SBT3..	SBb(rs)I−II		T=3	N15

The nuclear region dominates the entire system. Star formation activity is present in the bar and lens as well as in the faint spiral arms.

NGC 4535

779	6	79 03 25 08.0		McD 0.9m
B=10.6	B−V=0.70	U−B=0.05		V=1853
.SXS5..	SBc(s)I.3		T=5 L=1*	N08

The spiral arms appear to attach primarily at the leading edges of the short bar. The spiral pattern is unusually well developed, both in intermediate to old population stars and with respect to the distribution of blue knots, many of which are close to the leading edge of the arms. Plume structure is evident, particularly between the arms at the left.

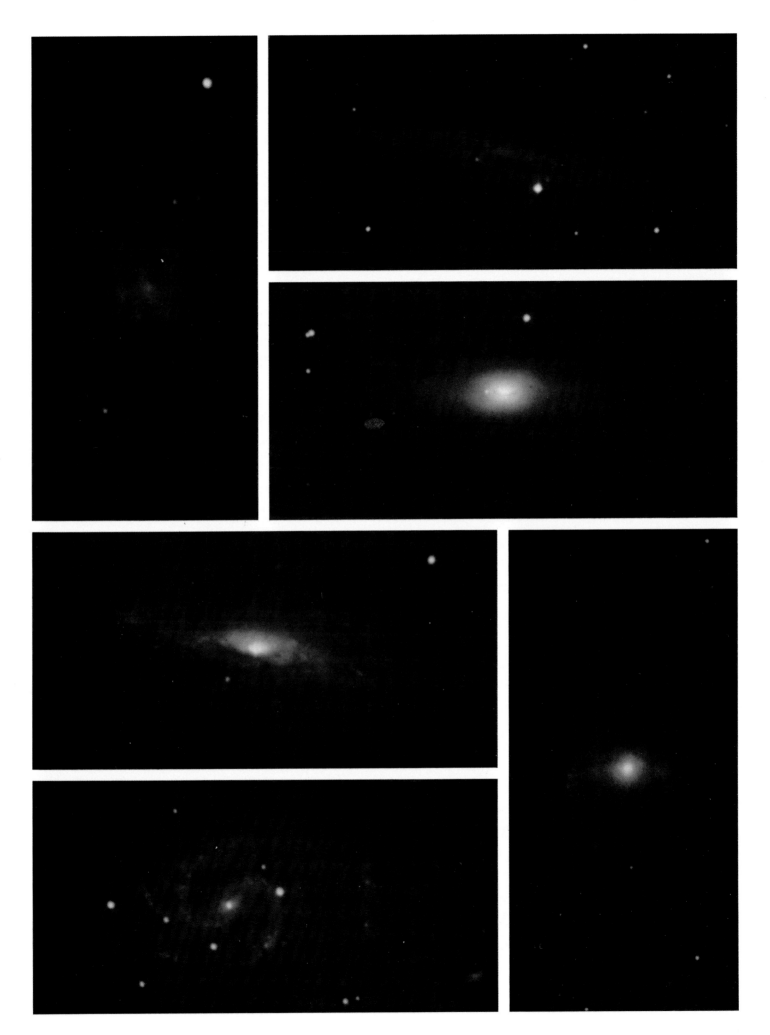

NGC 4550

1162	6	80 04 19 09.3		McD 0.7m
B=12.4	B−V=0.91	U−B=0.43		V=275
.LB.+$/	E7/S0₁(7)		T=−1	N06

Both NGC 4550, below left, and NGC 4551 appear to be comprised entirely of the old yellow stellar population.

NGC 4559

519	6	78 03 05 10.3		McD 0.7m
B=10.3	B−V=0.45			V=802
.SXT6..	Sc(s)II−III		T=6 L=4	N05

A weakly defined yellow bar is visible. The dust lanes exhibit the highest degree of organization in the system, particularly the one interior to the arm at the left. Star formation activity is widespread and extends beyond the bright intermediate aged disk region.

NGC 4564

1585	3	82 03 31 08.9		McD 2.7m
B=11.9	B−V=0.96	U−B=0.5		V=942
.E.6...	E6		T=−5	N18

This galaxy exhibits only an old yellow population. A hint of structure at the lower right may be associated with dust.

NGC 4565

1509	6	82 03 27 10.0		McD 0.7m
B=10.3	B−V=0.83			V=1122
.SAS3$/	Sb		T=3 L=1*	N23

The extraordinarily massive dust lane (actually a dust disk seen edge-on) fortuitously masks enough of the light of the nuclear region to reveal the point-like nucleus itself. There is no sign of star formation in the system, even in the visible outer disk, but it is possible that star formation may be present in the disk although hidden by dust.

NGC 4569   M 90

494	6	78 01 09 11.8		McD 0.7m
B=10.2	B−V=0.75	U−B=0.30		V=−382
.SXT2..	Sab(s)I−II		T=2	N09

This remarkable system has tightly wound smooth broad spiral arms. The arms are not yellow, as is the nuclear region, nor do they present the patchy appearance that accompanies embedded blue knots. Instead, it appears that this galaxy possess a complete 'fossil' spiral arm system (see NGC 4258). With the exception of the inner disk region the entire galaxy may have stopped producing stars, or at least the massive ones which give blue knots their color and brightness. As a consequence the system appears to be evolving in the direction of NGC 4826, probably on its way to becomming an S0 type galaxy.
A companion blue Magellanic irregular galaxy also present in the field is either less efficient at evolution or is a more recent formation than NGC 4569.

NGC 4579   M 58

1533	6	82 03 27 08.7		McD 0.7m
B=10.6	B−V=0.83	U−B=0.32		V=1730
.SXT3..	Sab(s)II		T=3	N18

As in NGC 4548 the nuclear region dominates the bar and the galaxy as a whole. The spiral structure is faint. A single blue knot is visible in the spiral arm region. Note the narrow yellow libration arcs just beyond the bar ends.

NGC 4586

930b	6	79 05.9		McD 0.9m
B=12.6	B−V=1.01			V=691
.SAS1*/	Sa		T=1 L=5	N18

Note the peculiar light distribution along the upper visible boundary of the galaxy near the nuclear region. There is a nuclear 'indentation' instead of a nuclear bulge. See also NGC 4192. Is it possible that this appearance is due to the same process responsible for nuclear 'box' structure in other galaxies? If so the end-on bar model appears to fail here because of the orientation of the dust lanes. Dust lanes associated with bars tend to parallel the bar axis, and when viewed in end-on projection they appear sharply inclined with respect to the disk as in NGC 4274. NGC 4192 is also similar in lacking evidence for the presence of a bar.

NGC 4593

1557	3	82 03 30 08.4		McD 2.7m
B=11.7*				V=2560
RSBT3..	SBb(rs)I−II		T=3 L=3	N16

A peculiar aspect of this galaxy is the rotation of the major axis of the inner bar with respect to the major axis of the large faint bar. The outer bar has an offset similar to that of NGC 4256. There is a hint of possible star formation activity in the bright nuclear region.

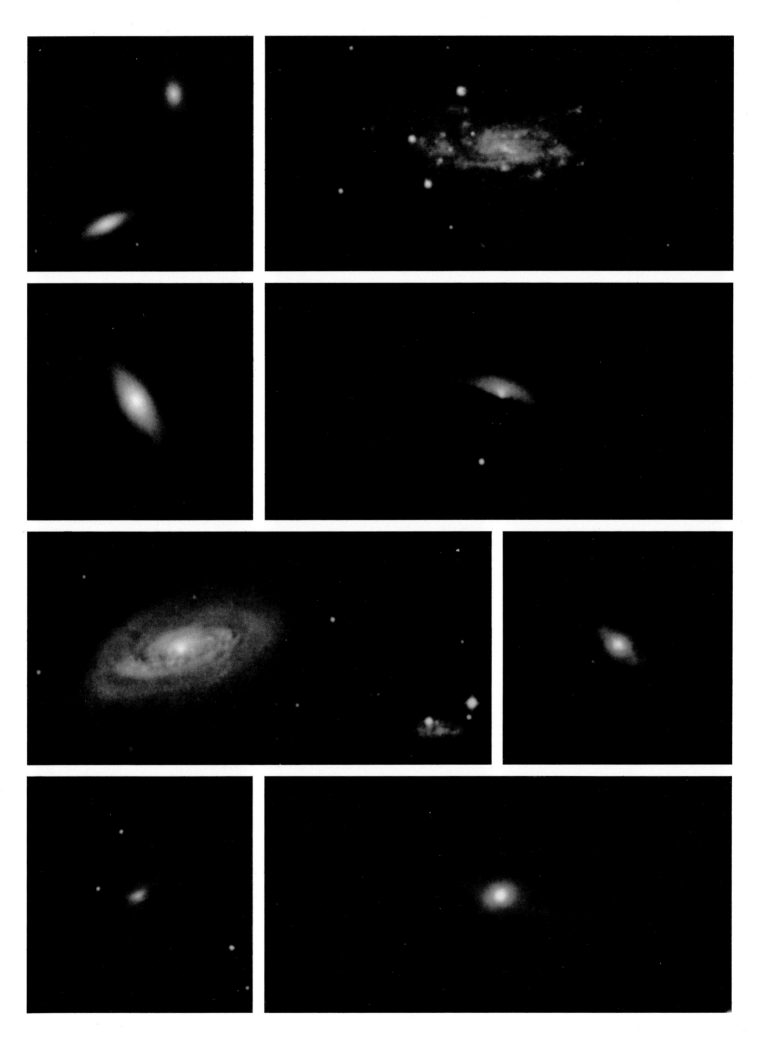

NGC 4596

836	3	79 04 25 07.2			McD 2.7m
B=11.4		B−V=.90	U−B=0.49		V=1789
.LBR+..		SBa(v early)		T=−1	N09

This highly evolved stellar system still retains the gap between bar end and (very faint) spiral arm, although each arm attaches to the bar at the opposite end. This configuration requires an evolutionary dymanics model based on more complex grounds than the relatively simple libration ring proposed for NGC 936 and others.

NGC 4594 M 104

238	6	76 04 01 04.1			McD 0.9m
B= 9.2		B−V=0.97	U−B=0.52		V=963
.SAS1/.		Sa+/Sb−		T=1	N00

This galaxy has an immense halo of old yellow stars. A suggestion of either intermediate aged or young stars is seen in the disk, but it is largely overwhelmed by the light of old stars together with reddening and extinction due to dust. The red spot at 12,1 is a defect. Corwin's photometry of the bright star gives V=10.12, B−V=0.63 and U−B=0.13.

NGC 4594 M 104

1200	6	81 05 01 05.2			CTIO 0.9m
B=9.2		B−V=0.97	U−B=0.52		V=963
.SAS1/.		Sa+/Sb−		T=1	N00

This sequence of three observations of NGC 4594 illustrates differences in properties of three of the instrumental systems (comprising telescope and detector combinations) used in the observing program for this atlas. The three individual detectors used in this program are represented here. The upper photograph (plate #238) was taken with the first of two RCA Carnegie two-stage image tubes used at the McDonald Observatory. Plate #1200 was taken with a similar image tube belonging to the Cerro Tololo Inter-American Observatory.

NGC 4597

1558	3	82 03 30 08.6			McD 2.7m
B=12.6*					V=905
.SBT9..		SBc(r)III:		T=9	N20

A faint yellow population and faint blue knots are visible in this very low surface brightness galaxy.

NGC 4594 M 104

1251	6	81 05 06 04.9			LC 1.0m
B=9.2		B−V=0.97	U−B=0.52		V=963
.SAS1/.		Sa+/Sb−		T=1	. N00

Plate #1251 and succeeding plates were made with the second McDonald Observatory Carnegie image tube. This image tube was used for the obsevations obtained at the Las Campanas Observatory in Chile.
Inspection of these three images reveals slight differences in quantum efficiency, but almost no difference in color response. The second McDonald image tube subsequently exhibited a reduction in ultraviolet and blue sensitivity, however, which resulted in the 1500 series plates being discernibly redder than the average color balance.
A number of globular clusters are visible in the field. The red object at 33,3 and the dark object at 05,2 are defects.

NGC 4605

1140	3	80 03 21 09.9			McD 2.1m
B=10.9					V=286
.SBS5P.		Sc(s)III		T=5	N20

This high surface brightness galaxy appears to have an old population component although it is dominated by intermediate aged and young stars. There is a peculiar asymmetry in the population distribution which gives the appearance of two distinct stellar systems: a system comprised primarily of young stars and chaotic appearing dust lines, approximately centered in the field with major axis nearly horizontal; and a system comprised of smoothly distributed intermediate to old aged (or metal-poor) stars centred at about 26,2 with major axis inclined about 20 degrees to the horizontal. This smooth system appears to lie behind the other.

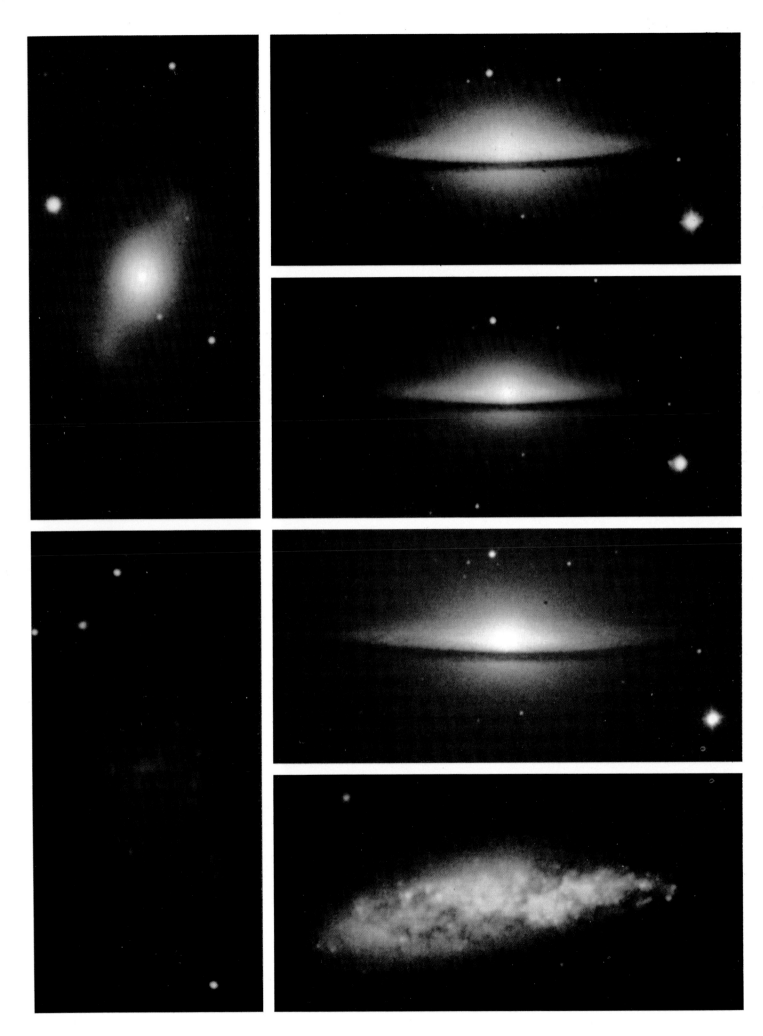

4608
4618
4621
4631
4627
4633
4634
4636
4643

NGC 4608

879 3 79 04 27 06.4			McD 2.7m
B=12.0 B−V=0.96			V=1721
.LBR0.. SB03/a		T=−2	N00

The nuclear region dominates the bar, which is the only part of the galaxy bright enough to record here. The population appears to be entirely old.

NGC 4618

226 6 76 03 31 09.1			McD 0.9m
B=11.2 B−V=0.43 U−B=−.15			V=613
.SBT9.. SBbc(rs)II.2p		T=9	N00

There is no suggestion of yellow in this system, and although the central region is overexposed, this is a possible candidate for an intrinsically young galaxy. Star formation activity is intense throughout the system. It is inconceivable for such a level of activity to persist over half of a Hubble Time without having produced a readily detectable old stellar population. Hence, if the system is old it must have been an inefficient producer of stars for much of its existence.

NGC 4621   M 59

1586 3 82 03 30 09.1			McD 2.7m
B=10.7 B−V=0.96 U−B=0.50			V=341
.E.5... E5		T=−5	N00

No young nor intermediate aged stars are evident in this system.

NGC 4631

269 6 76 04 01 07.9			McD 0.9m
B=9.7 B−V=0.54			V=638
.SBS7./ Sc on edge		T=7 L=5*	N32

This incredible galaxy is totally engulfed in star formation activity. A yellow population is detectable in the nuclear region. Dust is present, but not in massive amounts, nor confined to a thin disk.

The companion galaxy, NGC4627, is the prototype metal-poor dwarf elliptical for the atlas. Its photometric colors are B−V(T)=0.62 and U−B(T)=0.13, with which the apparent color agrees very closely. If this galaxy is assumed to be as old as the old yellow stellar population of other galaxies then the greenish color does not uniquely indicate intermediate age. In most instances in the atlas, however, the green color present in the disk of a system containing a yellow (hence metal-rich) nuclear region may be assumed to belong to a population that is at least as metal-rich as the nuclear region, and hence the color is unambiguously due to intermediate age. In those cases where no yellow population is visible the question of interpretation of a green appearing stellar population becomes cosmologically important. Did all galaxies form at about the same time after an initial 'Big-Bang', or did some galaxies, for whatever reason, form at relatively recent epochs? If galaxy formation occurs over an extended period of time, then the models predicting galaxy formation at a single critical epoch after the Big-Bang would need to be either reevaluated or discarded altogether in favor of models permitting galaxy formation to take place over the relevant extended period. It is not possible within the context of these observations to cite an unambiguous case which directly confirms recent galaxy formation, but it is possible to isolate candidates for further investigation in which the finding of normal or high metallicity would confirm their recent formation. See also the discussion accompanying NGC1313 and the galaxies referred to therein.

NGC 4636

1588 3 82 03 31 10.0			McD 2.7m
B=10.5 B−V=0.94 U−B=0.50			V=869
.E.0+.. E0/S01(6)		T=−5	N00

The globular cluster population appears to be just detectable in this photograph. No young or intermediate aged stars are visible. Another galaxy is seen at 27,3.

NGC 4633

1164 6 80 04 19 09.8			McD 0.7m
B=13.8 B−V=0.65			V=209
.SXS7*.		T=7	N04

NGC 4633 is the faint blue galaxy barely visible at the upper right. The edge-on galaxy at the lower left is NGC 4634.

NGC 4643

880 3 79 04 27 06.5			McD 2.7m
B=11.5 B−V=0.95 U−B=0.55			V=1234
.SBT0.. SB03/SBa		T=0	N00

This system comprised of old stars appears similar to NGC 4608 above. If the perspective corresponds to that of NGC 1398 then there is a noticable absence of libration arc structure.

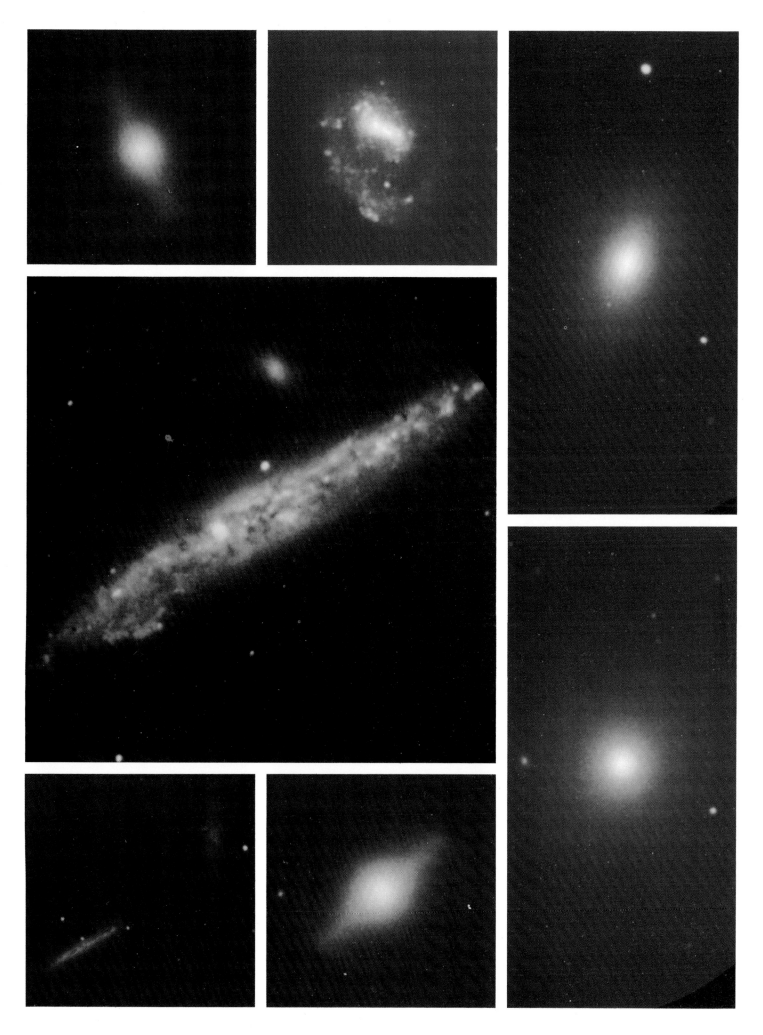

4649
4647
4651
4654
4656
4657
4665

NGC 4649   M 60

1562	3	82 03 30 09.3	·	McD 2.7m
B=9.8	B−V=1.00	U−B=0.55		V=1128
.E.2...	SO1(2)		T=−5	N31

NGC 4649 (below) is comprised of an old
yellow stellar population. NGC 4647 has a
small yellow bar surrounded by an inner ring
defined largely by regions of active star
formation. The bar contains star formation
activity also. The outer disk is patchy, blue and
does not exhibit spiral structure.

NGC 4651

1093	6	80 03 19 08.7		McD 2.1m
B=11.3	B−V=0.58			V=742
.SAT5..	Sc(r)I−II		T=5 L=3*	N00

The broad yellow disk appears to drop off
rapidly in surface brightness. The color of the
disk masks the color of blue knots in the inner
ring. Spiral arms outside the disk are faint but
well defined by the intermediate population.
Blue knots are present in the spiral arms, but
star formation activity appears to be greater in
the inner disk.

NGC 4654

933b	6	79 05.9		McD 0.9m
B=11.1	B−V=0.64	U−B=−.07		V=970
.SXT6..	SBc(rs)II		T=6 L=3	N19

The inner region and bar appear reddened.
Compare with NGC 4559. Star formation is
confined to inner and mid-disk regions. Blue
knots in the outer disk are few and very faint.
Corwin's photometry of the red star is
V=9.55, B−V=1.32 and U−B=1.46.

NGC 4665

1559	3	82 03 30 08.7		McD 2.7m
B=11.4*				V= 678
.SBS0..	SB01/3 SBa		T=0	N21

Only an old yellow population is visible here.
Corwin's photometry of the bright star is
V=10.84, B−V=0.66 and U−B=0.12.

NGC 4656

1510	6	82 03 27 10.5		McD 0.7m
B=10.7	B−V=0.43			V=662
.SBS9P.	Im		T=9 L=7	N12

The central region of the galaxy is at the center
of the photograph. The left side of the galaxy is
too faint to record here. There is a definite
population gradient from green to blue from
the center outward along the arm to the right.

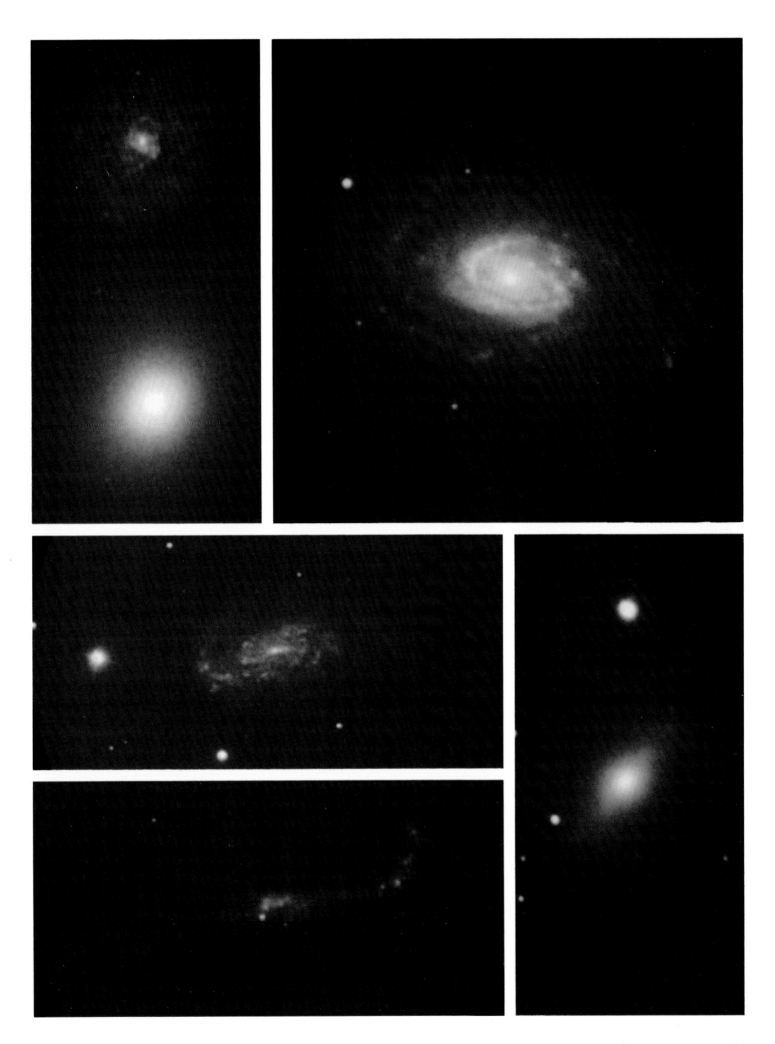

4666
4676
4697
4699
4710
4725

NGC 4666

1277 6 81 05 07 04.4			LC 1.0m
B=11.5	B−V=0.79	U−B=0.23	V=1395
.SX.5*.	SbcII.3	T=5	L=2* N30

The yellow central region is surrounded by dense dust lanes in the disk. The disk is of mixed population comprising old, intermediate aged, and young stars.

NGC 4710

882 3 79 04 27 06.8			McD 2.7m
B=11.8	B−V=0.83	U−B=0.25	V=1076
.LAR+$/	S03(9)	T=−1	N10

Here we have the remarkable combination of an old yellow stellar population, a box shaped nuclear region, an inner dust lane and star formation activity in the dust lane. Star formation activity in highly evolved (yellow) stellar systems is not entirely uncommon, however. See for example NGC 3626 and especially NGC 4826 which may be quite similar to NGC 4710.

NGC 4697

1252 6 81 05 06 05.2			LC 1.0m
B=10.2	B−V=0.93	U−B=0.39	V=1170
.E.6...	E6	T=−5	N02

A pure yellow population is seen here.

NGC 4699

1278 6 81 05 07 04.7			LC 1.0m
B=10.4	B−V=0.88	U−B=0.34	V=1359
.SXT3..	Sab(sr)	T=3	N18

The disk is patchy and partially reddened by dust. There is a suggestion of green indicating the presence of intermediate or possibly young stars in a broad ring at the edge of the bright inner disk.

NGC 4725

269 6 76 04 01 09.0			McD 0.9m
B=9.9	B−V=0.74	U−B=0.23	V=1131
.SXR2P.	Sb/SBb(r)II	T=2	L=1 N00

The nuclear region dominates the bar structure as well as the rest of the galaxy. The bar is very broad, but quite faint. There is no evidence of star formation interior to the surrounding ring of intermediate population and blue knots. Some smooth possibly yellow spiral features are seen in the ring region near 30,3.

NGC 4676

833 3 . 79 04 25 06.5			McD 2.7m
			V=6631
.L...P$		T=−2	N00

NGC 4676a (above) and NGC 4676b form a close interacting pair. Both galaxies have blue knots at the ends of their major axes. NGC 4676b in particular is somewhat similar to NGC 6054 and NGC 3561a (also in an interacting system). NGC 4151, a galaxy with an active nucleus, also has blue bar ends although they are much fainter than any of the above galaxies. It is tempting to suggest that these occurrences are not simply coincidental, but rather that they are based on active mass-flow processes, probably originating in the nuclear regions, which are triggered into activity or otherwise enhanced as a result of tidal encounters. Certainly the possibility of nuclear activity as a cause of these peculiar blue bar-end enhancements deserves further consideration.

The morphological nature of the inner region of NGC 4676a is nothing less than baffling.

NGC 4731

1237 6 81 05 04 03.3   CTIO 0.9m
B=11.4   B−V=0.40   U−B=−.25   V=1351
.SBS6..   SBc(s)III: T=6 L=4* N09

This system has a narrow blue bar leading into broad, well defined spiral arms. A bright blue knot near the bar together with a faint arm segment bifurcating from the main arm at 18,2 may be indicators of a weak and disorganized inner ring structure. There is some question here of whether or not this system has any old population stars.

NGC 4736   M 94

223 6 76 03 31 08.7   McD 0.9m
B=8.8   B−V=0.75   U−B=0.16   V=329
RSAR2..   RSab(s)   T=2 L=3* N18

The inner disk is extremely bright, and is also illustrated below in a short exposure image. The outer disk is rather yellow and generally smooth. Blue knots of star formation are present along two barely discernible spiral arms which wind outward through the disk; the density wave eking out the last bits of star forming material as the outer disk prepares to enter its final evolutionary stage. See NGC 4826 and below.

NGC 4753

1560 3 82 03 30 08.9   McD 2.7m
B=10.8   B−V=0.95   U−B=0.48   V=1137
.I.0...   S0p   T=0   N09

This is one of a small class of galaxies of the NGC 4438 type where distortion of the disk in the vertical direction brings dust patterns in the disk into projection against the nuclear region.

NGC 4736   M 94

1104 0.5 80 03 19 11.8   McD 2.1m
B=8.8   B−V=0.75   U−B=0.16   V=329
RSAR2..   RSab(s)   T=2 L=3* N18

The inner disk remains extremely active, in marked contrast to the outer. In this short exposure at higher scale we see the bright nuclear region surrounded by an inner disk of old stars mixed with a major complex of dust lanes surrounded in turn by a blue ring of intense star formation activity. Dust lanes winding out from the nuclear region break this ring at 06,2 and 23,2, at which points the bright ring appears linked to the faint arms visible in the upper photograph. It seems clear that the real dynamics driving the star formation here is related to processes associated with the dust lanes, and therefore quite likely the mechanism is associated with the origin of the dust lanes within the nuclear region.

NGC 4754

1095 6 80 03 19 09.5   McD 2.1m
B=11.4   B−V=0.95   U−B=0.50   V=1393
.LBR−*.   S0p   T=−3   N00

This system gives the appearance of a bar seen in projection from an angle slightly off the major axis of the bar. Diffuse appendages at the upper left and lower right suggest residuals of spiral features.

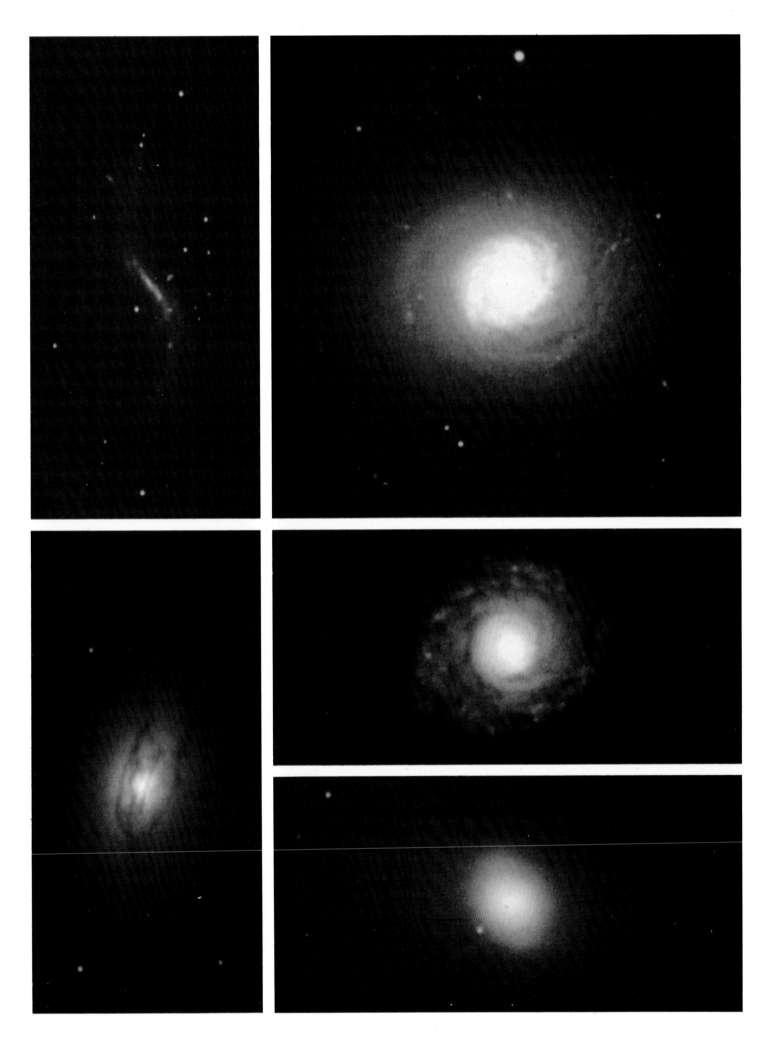

4818
4826
4900
4902
4921
4945

NGC 4818

1279 6 81 05 07 05.0				LC 1.0m
B=11.9*				V=958
.SXS2..	Sab:		T=2	N05

There is little suggestion of current star formation activity in this system, although it contains an intermediate aged population as well as an old one. A weakly defined spiral structure is present.

NGC 4900

726 6 79 03 22 08.3				McD 0.9m
B=12.1	B−V=0.60	U−B=−.15		V=945
.SBT5..	ScIII−IV		T=5 L=5	N00

A short bar is visible. Some plume structure is evident, but spiral structure is mostly lacking. This galaxy is similar to NGC 3226 except for its higher surface brightness. There is considerable star formation activity, but the disk is quite green.
Corwin's photometry of the star is V=11.31, B−V=1.29 and U−B=1.37.

NGC 4902

1280 6 81 05 07 05.3				LC 1.0m
B=11.9	B−V=0.74	U−B=0.09		V=2564
.SBR3..	SBb(s)I−II		T=3 L=1	N00

The yellow bar is surrounded by a blue ring. The outer disk is a complex of relatively smooth spiral features mostly lacking in star formation activity.

NGC 4921

1141 6 80 03 21 10.2				McD 2.1m
B=13.0	B−V=0.89	U−B=0.35		V=5468
.SBT2..			T=2	N00

The bright yellow bar dominates this galaxy. Only a few blue knots are bright enough in this photograph to provide visible evidence of outer spiral structure.

NGC 4826   M 64

1564 3 82 03 30 09.7				McD 2.7m
B=9.3	B−V=0.84	U−B=0.21		V=377
RSAT2..	Sab(s)II		T=2	N20

This is the prototype example for the class of galaxy comprising a highly evolved system with second wave star formation activity; an Evolved Second Wave Activity Galaxy (ESWAG).
Most galaxies with active star formation exhibit a blue gradient; i.e. the star formation activity is greatest near the outside of the visible disk. Since the center of the disk is normally old (yellow) we infer that star formation begins most strongly in the center and then works its way outward over a period

NGC 4826   M 64

90 10 75 05 04 07.7				McD 0.9m
B=9.3	B−V=0.84	U−B=0.21		V=377
RSAT2..	Sab(s)II		T=2	N20

The 1500 series photographs (example above) are slightly redder in general than the average atlas color balance. In this case the difference is small.

NGC 4945

1239 6 81 05 04 03.9				CTIO 0.9m
B=9.4				V=356
.SBS6*/	Sc		T=6	N27

Only thirteen degrees from the plane of the Milky Way this object is both self reddened and reddened by dust in our galaxy. In addition to the complex almost radial pattern of dust notice the bright segment of contiguous blue knots at the edge of the disk at the upper right. Such alignments of blue knots are relatively common over a wide range of galaxy types. Evidently conditions relating to star formation can be quite similar over large scale galactic dimensions, yet are subject to abrupt variations. See also NGC 1532 for example. This galaxy is similar to NGC 253.

of typically several billion years. This process is basically one of digesting the star forming materials in order of decreasing abundance. A number of systems, however, exhibit star formation activity in the central regions (usually in a ring-like structure) as star formation acitivty in the outer regions is apparently dying out (NGC 4736 and NGC 4569). Others, like NGC 4313, NGC 4710, NGC 3626 and NGC 4826 have interior star formation activity long after star formation activity has died out in the outer regions. The two waves of star formation activity thus seem to be fundamentally different in nature. In particular, the material feeding the first wave of star formation is depleted generally from the interior outward, so that eventually only the outer regions have enough interstellar material to form stars under the influence of either density waves or stochastic processes. How then can the second wave of star formation occur? Although in-fall of gas from an intergalactic medium may be a possibility, most likely the activity is fueled by materials generated in the interior of the galaxy where the second wave originates. As stars evolve, they lose mass, gradually restoring the interstellar medium. In regions of high stellar density (i.e. the nuclear region) the interstellar medium increases in density most rapidly, until the material given off by the aging stars becomes sufficient to support a second wave of star formation.

Two principal factors cloud this picture: What mechanism feeds the shock front with fresh material as it works its way outward; and why are ESWAGs limited to flattened (rotating) stellar systems? I suspect that one place to look for part of the answers to these questions is into the nature of active nuclei in galaxies. It should be borne in mind that the original interstellar medium is not depleted to zero density by the first wave of star formation, but simply to a level insufficient to support star formation activity. Consequently the amount of recycled material necessary to initiate the onset of a second wave is considerably less than the amount originally present in the galaxy.

Finally, star formation activity in the nuclear regions of many galaxies indicates that second wave conditions are established in many galaxies even long before the first wave of star formation has begun to die out in the outer regions, as for example in NGC 4321. Although these are not highly evolved systems they are nevertheless second wave (star formation) activity galaxies (SWAGs).

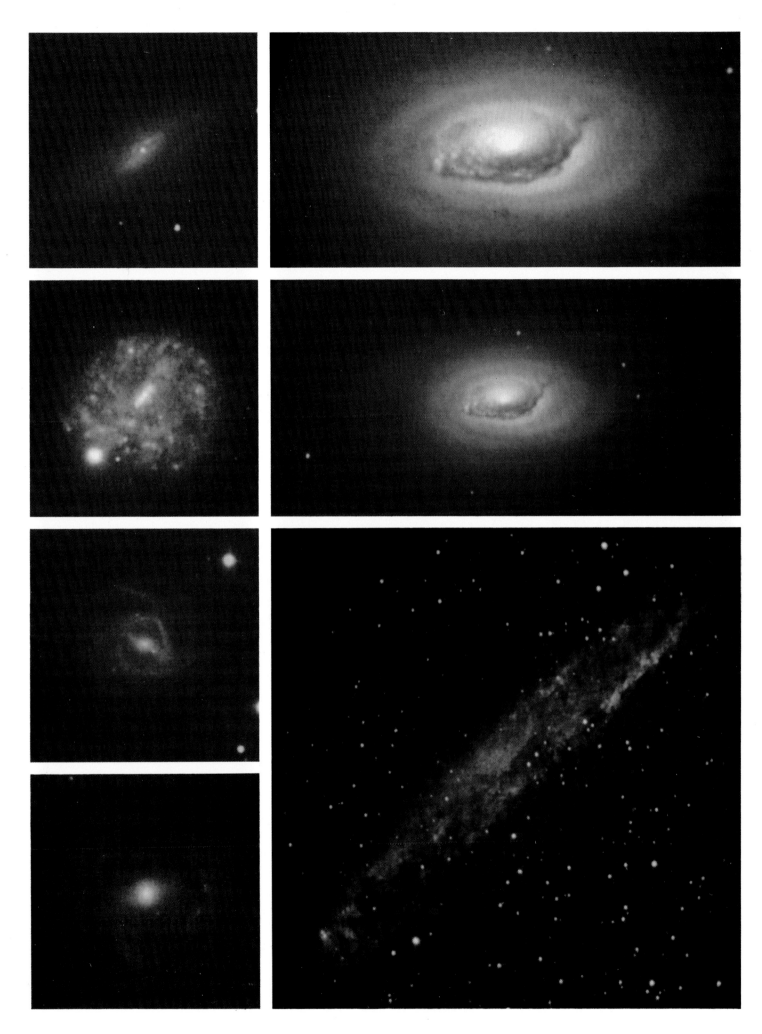

69

4965
4976
5005
5033
5054
5055

NGC 4965

1281	6	81 05 07 05.6			LC 1.0m
					V=2073
.SAT8..			T=8		N32

This galaxy is faintly visible over much of the field, but its surface brightness is so low that only the yellow nuclear region and several small blue knots are reasonably discernible here.

NGC 5005

1565	3	82 03 30 09.8			McD 2.7m
B=10.6	B−V=0.82		U−B=0.30		V=1069
.SXT4..	Sb(s)II			T=4 L=3	N15

Dust lanes emanating from the bright nuclear region establish the spiral pattern for the galaxy. Many star formation regions are associated with plumes. The disk contains a large yellow population which, although patchy, is much smoother than that of NGC 5055 (below) for example. Compare with NGC 4736.

NGC 5033

781	6	79 03 25 08.6		McD 0.9m
B=10.6	B−V=0.54			V=961
.SAS5..	Sbc(s)I−II		T=5 L=2*	N09

The disk immediately surrounding the yellow nuclear region is comprised of blue knots, regions of intermediate color and numerous dust lanes. Spiral structure defined by faint blue knots (one at 18,2) extends far beyond this inner disk.

NGC 4976

1240	6	81 05 04 04.1		CTIO 0.9m
B=11.1	B−V=0.97	U−B=0.33		V=1133
.E.4.P*	S0₁(4)		T=−5	N09

This galaxy appears to be comprised entirely of an old yellow population.
This southern hemisphere field was not accessible for photometry by Corwin.

NGC 5054

1282	6	81 05 07 05.9		LC 1.0m
B=11.5	B−V=0.80	U−B=0.02		V=1594
.SAS4..	`Sb(s)I−II		T=4 L=3*	N24

The small yellow nucleus appears to possibly contain star formation activity. The disk has a spiral pattern but only one bright blue knot. Most of the disk is probably of intermediate age.
A blue galaxy is visible at 26,5.

NGC 5055   M 63

758	6	79 03 24 09.3		McD 0.9m
B=9.3	B−V=0.73			V=587
.SAT4..	Sbc(s)II−III		T=4 L=3	N00

Here the patchy pattern often associated with lack of spiral structure is strongly influenced by an overall plume structure with characteristic spiral pattern. Yellow light dominates the inner region while blue dominates the outer. A number of bright compact blue knots are present.
Corwin's photometry of the bright star is V=9.30, B−V=0.47 and U−B=0.01.

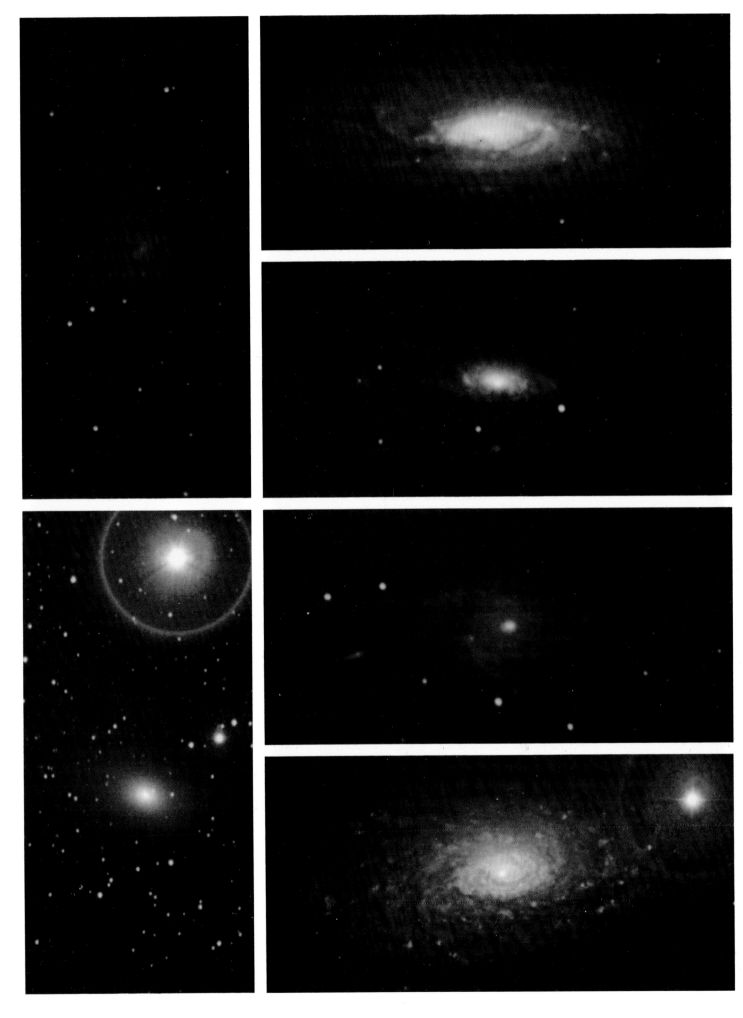

5068
5085
5101
5102
5112

NGC 5068

1222	6	81 05 02 05.0		CTIO 0.9m
B=10.2	B−V=0.60	U−B=−.20		V=402
.SXT6..	SBc(s)II−III		T=6	L=6 N09

A blue bar with white nuclear region is normal for very late-type systems such as this. See also NGC 4731, NGC 3664 and particularly NGC 3319. This system appears to possibly contain old yellow stars. If one such system can be shown to contain old stars then similar systems may reasonably be assumed to be old also. See NGC 5112 below.

NGC 5085

1253	6	81 05 06 05.5	LC 1.0m
B=11.9*			V=1783
.SAS5..	Sc(r)I−II	T=5	L=4 N18

The small yellow nucleus is surrounded by a smooth intermediate color disk containing several faint blue knots.
Corwin's photometry of the bright star is V=8.44, B−V=1.23 and U−B=1.27

NGC 5101

1254	6	81 05 06 05.8		LC 1.0m
B=11.0	B−V=0.95	U−B=0.50		V=1675
RSBT0..	SBa		T=0	N16

This galaxy appears quite similar to NGC 4596. The red and blue spots on the lower left part of the bar are plate defects.

NGC 5102

1243	6	81 05 04 04.9		CTIO 0.9m
B=10.3	B−V=0.70	U−B=0.27		V=247
.LA.−..	S0$_1$(5)		T=−3	N31

Only the old population is evident in this galaxy. The photoelectric B−V appears to be anomalous. The field contains a number of foreground stars.

NGC 5112

1566	3	82 03 30 10.0		McD 2.7m
B=11.8*				V=1030
.SBT6..	Sc(rs)II		T=6	N04

The nuclear region appears yellow. It is flanked by two green knots (see NGC 3319 and NGC 5068). If these are blue knots then dust reddening might explain the color of the nuclear region. More likely the color is in fact indicative of an old yellow population.

1295 6 81 05 08 03.6		LC 1.0m
B=7.9 B−V=0.98		V=323
.L...P. S0+Sp	T=−2	N12

One of the most remarkable galaxies known, a
radio source, an X-ray source, an object with
star formation occurring out to enormous
(radio lobe scale) distances and
morphologically incredible on the galactic
scale seen here.

The central yellow system appears nearly
spherical. It appears to be surounded by a
tremendously massive band of dust. The light
from the nuclear region is bright enough to
shine through this dust, becoming reddened in
the process by half a magnitude or more.
Shards of dust are evident high above the main
plane of the dust. Embedded in the dense dust,
particularly along its lower edge, are numerous
blue knots, regions of active star formation.
There is no definite evidence for an
intermediate aged stellar population. At the
upper left a thin veil of dust, sharply bounded
(the edge passes near the bright white star), is
relatively smooth over what appears to be
galactic dimensions.

Slightly dark red and yellow color defects are
pesent near 32,7. The photograph is a
composite of two sets of exposures.

NGC 5161

1244 6 81 05 04 05.1			CTIO 0.9m
B=12.0*	B−V=0.80	U−B=0.10	V=2018
.SAS5*.	Sc(s)I		T=5 L=7* N30

The system has a poorly defined inner ring, but little evidence for an associated bar. Several blue knots are present in the outer spiral region.

NGC 5170

1294 6 81 05 08 03.1			LC 1.0m
B=12.0	B−V=0.85	U−B=0.17	V=1350
.SAS5*/	Sb:		T=5 N00

The nuclear region appears to have a box structure. The disk contains massive amounts of dust. A star formation region is visible at 25,3. See also NGC 4216.

NGC 5172

1096 6 80 03 19 09.8			McD 2.1m
B=12.6	B−V=0.73	U−B=0.10	V=4014
.SXT4*.	SbI		T=4 L=2 N00

The nuclear region is yellow as is a large area of the inner disk. A long smooth section of spiral arm in the inner disk is mostly lacking in star formation activity. In the outer disk one spiral arm segment is sharply defined by blue knots (see also NGC 772), while another comprises a broad region of plume structure.

NGC 5194   M 51

1102 0.5 80 03 19 11.5			McD 2.1m
B=8.9	B−V=0.60	U−B=−.06	V=565
.SAS4P.	Sbc(s)I−II		T=4 L=1 N17

This is a very short exposure of the nuclear region of NGC 5194. It shows a ring of numerous star formation regions surrounding a bright yellow inner disk and nucleus. See below.

NGC 5195   M 51

1105 3 80 03 19 11.9			McD 2.1m
B=10.5	B−V=0.90	U−B=0.40	V=658
.I.0.P.	SB01p		T=0 N18

Dust from NGC 5194 partially obscures and reddens the light from this galaxy. The blue knots at the top and also immediately to the right of NGC 5195 are associated with spiral features in NGC 5194. Only an old yellow stellar population is evident in NGC 5195.

NGC 5194   M 51

1101 1.5 80 03 19 11.4			McD 2.1m
B=8.9	B−V=0.60	U−B=−.06	V=565
.SAS4P.	Sbc(s)I−II		T=4 L=1 N17

This short exposure is three times the duration of the one above and shows the inner pattern of plume structures surrounding the nuclear region. Several short smooth yellow arm segments are visible at the top edge of the inner disk, and also at its lower left. These appear to be comprised of old population stars. These may represent a multiple wave pattern encircling the nuclear region. The remaining spiral features are mostly dominated by blue knots and associated star formation activity.

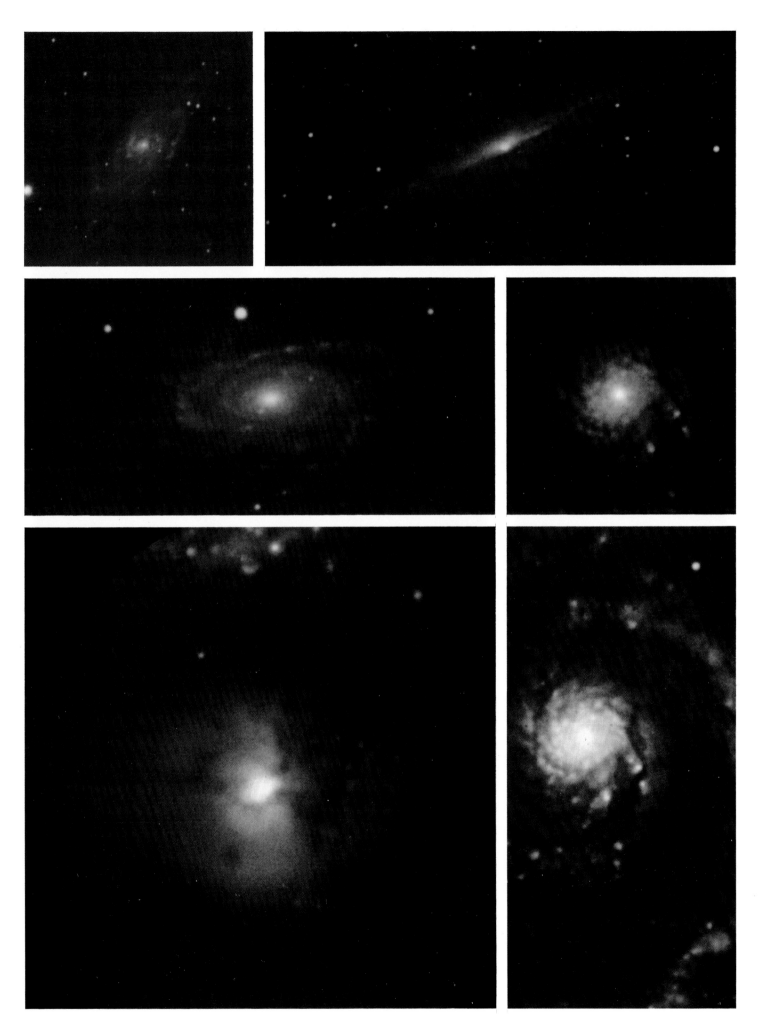

NGC 5194   M 51

235   6   76 03 31 11.2   McD 0.9m
B=8.9   B−V=0.60   U−B=−.06   V=565
.SAS4P.   Sbc(s)I−II   T=4 L=1 N17

NGC 5194 and NGC 5195 form an interacting pair. The upper arm of NGC 5194 extends down on the right all the way across the line of sight to NGC 5195.

The inner disk of NGC 5194 contains an old population visible here in the inter-arm areas. Notice in the bright arm on the right two distinct zones of different color: the upper one being a greenish blue and relatively smooth, while the one at lower right is very blue and patchy with numerous blue knots. These differences in appearance indicate a marked difference in star formation acitivty in these two regions. In the upper region, if star formation is prevalent at all, it is restricted to production of lower mass stars. Clearly high mass stars are being produced in the lower arm region. Yet the density wave characteristics, the trailing dust lane and the older stars along the inner edges of the arm, appear to be well established and functioning in both regions, and certainly give no reason *a priori* to expect such differences between the two regions. Some factor is effective here in producing two significantly different results from apparently identical processes. One possibly significant difference is evident: the upper arm segment is radially interior to the lower one. Hence it is possible that star formation activity in general has become reduced in that region and the upper arm segment is in the process of becoming a fossil arm structure. See for example NGC 4258 and the discussion accompanying NGC 4826.

The inter-arm regions are dominated by plume structures. Some of these plumes clearly originate at bright blue knots, such as the relatively bright and narrow plume at 27,5. Another very prominent plume structure occurs at the gap at 11,6 between the two blue population groups. Do these plumes represent high frequency harmonics in the density wave? Do they represent the locus of differential orbital motion for stars and material drifting away from regions of active star formation? Whatever their cause, they are responsible for much of the pattern structure visible in spiral galaxies.

NGC 5204

810	6	79 03 26 11.0		McD 0.9m
B=11.7	B−V=0.49	U−B=−.15		V=351
.SAS9..	SdIV		T=9	L=7 N03

There is little indication of spiral pattern visible here. An old yellow population is present as are regions of star formation activity. The violet knot at 04,3 is probably associated with a luminous O star. Closer examination reveals that supergiants are resolved in this galaxy, giving it an informal uncorrected distance modulus estimate of about +28. See NGC 604 and NGC 625.

NGC 5230

1097	6	80 03 19 10.1		McD 2.1m
B=12.8*				V=6814
.SAS5..	Sc(s)I		T=5	L=2 N03

The small yellow nuclear region is surrounded by a partial blue ring. Massive long plumes extend from 30,1 to the arm defining the lower outer boundary. Another galaxy is visible at 30,3.

NGC 5236   M 83

1298	6	81 05 08 04.4		LC 1.0m
B=8.2	B−V=0.66	U−B=0.03		V=337
.SXS5..	SBc(s)II		T=5	L=2 N32

This galaxy is completely dominated by plume structure. Spiral features are associated with major dust lanes. Note in the bar on the right where many fine dust lanes cut the bar at right angles to the primary bar dust lane. The nuclear region appears to contain bright blue knots.

Supergiant stars appear to be resolved in this galaxy giving by visual inspection an informal approximate uncorrected distance modulus of about +28. See NGC 604 and NGC 625. If NGC 5236 and NGC 5204 are at nearly the same distance, then NGC 5236 is much more luminous than NGC 5204. This is in qualitative accord with their magnitude and luminosity class data.

NGC 5247

1258	6	81 05 06 07.1		LC 1.0m
B=11.1	B−V=0.59	U−B=−.10		V=1511
.SAS4..	Sc(s)I−II		T=4	L=2 N04

Blue star formation regions are present immediately adjacent to the yellow nuclear region. Faint spiral arms form a well defined spiral pattern.

NGC 5248

762	6	79 03 24 10.4		McD 0.9m
B=10.8	B−V=0.63	U−B=0.00		V=1102
.SXT4..	Sbc(s)I−II		T=4	L=1 N04

The distribution of yellow light suggests the presence of a bar, and the inner spiral pattern is rather similar to that of NGC 5236 if allowance for the different projection perspective is made. See also the short exposure below.

NGC 5248

1088	0.5	80 03 18 11.3		McD 2.1m
B=10.8	B−V=0.63	U−B=0.00		V=1102
.SXT4..	Sbc(s)I−II		T=4	L=1 N04

This very short exposure at greater scale shows a ring of blue knots immediately surrounding the nucleus of NGC 5248. See also NGC 4321 and the discussion accompanying NGC 4826.

NGC 5257

838	3	79 04 25 07.6		McD 2.7m
				V=6791
.SXS3P.			T=3	N00

Although NGC 5257 on the right has a great deal of star formation activity only a few blue knots are visible in NGC 5258 on the left. An interaction between these two galaxies may have helped to trigger star formation in sensitive regions. Note the tidal distortion of the smooth green arm of NGC 5258.

NGC 5266

1286	6	81 05 07 07.1		LC 1.0m
B=12.3*				V=2923
.SA.6*.		S03(5)p(prolate)	T=6	N29

This galaxy is the prototype prolate spheroidal with a minor axis dust band illustrated in the atlas. Several other galaxies are suspect, including NGC 1316 and NGC 2768, but this is the only example with a continuous dust lane across the minor axis.

NGC 5300

1567	3	82 03 30 10.2		McD 2.7m
B=11.9*				V=1084
.SXR5..		/Sc(s)II	T=5	N18

The yellow nuclear region and a solitary stellar appearing bright blue knot punctuate this otherwise very faint spiral.

NGC 5301

1142	6	80 03 21 10.5		McD 2.1m
B=12.7*				V=1673
.SAS3*/		Sc(s)	T=3	L=4* N22

This galaxy appears to be divided into three principal regions: the yellow inner disk, a mid-disk region dominated largely by dust and an outer disk region dominated by stars of intermediate age with a number of blue knots.

NGC 5363

1098	6	80 03 19 10.4			McD 2.1m
B=11.2	B−V=0.99	U−B=0.58			V=1081
.I.0.$.	[S03(5)]		T=0		N05

Dust lanes cross the center of this yellow galaxy, sharply dimming the light from the nuclear region and making visible the area immediately surrounding the nucleus. The two red spots are defects.

NGC 5371

713	6	79 02 22 12.0			McD 0.9m
B=11.4	B−V=0.65	U−B=0.15			V=2660
.SXT4..	Sb(rs)I/SBb(rs)I		T=4	L=1	N22

A small but clearly defined yellow bar lies interior to a broad disk of well developed spiral structure comprising intermediate and young stellar populations. A singularly bright blue knot stands out in a narrow and sharply delineated string of star formation regions defining the upper arm.
Corwin's photometry of the bright star is V=8.93, B−V=1.23, and U−B=1.29.

NGC 5364

273	6	76 04 01 09.5			McD 0.9m
B=11.0	B−V=0.65	U−B=0.03			V=1349
.SAT4P.	Sc(r)I		T=4	L=1	N12

The entire spiral pattern has the appearance of partially rotated non-confocal ellipses. Note the bright blue knots in the lower arm near 16,3

5377
5383
5394
5395
5421
5427
5426
5448

NGC 5377

1143	6	80 03 21 10.8		McD 2.1m
B=12.0	B−V=0.82	U−B=0.30		V=1950
RSBS1..	SBa or Sa		T=1	N12

The right and left extremities appear to be bar-end regions with star formation acitvity as in NGC 4151. A foreground dust lane associated with the right bar is visible to the lower right of the nucleus. A brightening to the upper left of the nucleus may be the corresponding far arm of a diffuse central spiral pattern associated with the bar dust lanes, as in NGC 4314 but larger.

NGC 5394

839	3	79 04 25 07.8		McD 2.7m
B=13.6	B−V=0.66	U−B=0.12		V=3496
.SBS3P.			T=3	N06

NGC 5394 is the smaller exceedingly peculiar galaxy. The spiral arcs are extremely bright, yet they are mostly smooth and green. It appears that a massive burst of star formation, perhaps as in NGC 5257, was recently abruptly terminated in this galaxy and the resulting fossil arms now comprise the dominant morphological structure of the galaxy.

NGC 5395 on the left is distorted with several long dust lanes apparently cutting across most of the galaxy. The upper left arm is particularly bright, but possibly lacking in massive star formation activity, somewhat analogous to the blue knot poor arm section in NGC 5194. This arm may also be on the threshold of becoming a bright fossil spiral feature. See also the interacting pair NGC 2936–7.

NGC 5421

886	3	79 04 27 07.5	McD 2.7m
			V=7999 .
			N27

The blue irregular galaxy on the right may be a shard of gas torn free during a tidal interaction, which is now forming stars. If so it could be considered an intrinsically young galaxy, but its existence under such circumstances obviously would not be a factor in testing cosmological models for the origin of galaxies in general.
The spiral arms of the barred spiral on the left are sharply defined by star formation acitvity.

NGC 5448

1144	6	80 03 21 11.1		McD 2.1m
B=12.2				V=2100
RSXR1..	Sa(s)		T=1 L=4	N00

The yellow population interior to the ring appears to contain a bar. The brightest regions in the ring are in positions which correspond to bar-end enhancements. Note the narrow arc of almost contiguous blue knots at the left. This system is remarkably similar to NGC 3504, including the manner in which very faint outer arms attach to the bright, and evidently elongated, disk region.

NGC 5383

782	6	79 03 25 09.0		McD 0.9m
B=12.0	B−V=0.64	U−B=0.03		V=2354
.SBT3*P	SBb(s)I–II		T=3 L=3	N00

Two blue knots lie to either side of the small yellow nucleus. Star formation is occurring in at least one of the nuclear region 'arm' segments. The bar dust lanes are very massive. The leading edge of the bar appears to be blue as do the spiral features.

NGC 5427

1591	3	82 03 31 10.6		McD 2.7m
B=12.0	B−V=0.62	U−B=−.09		V=2483
.SAS5P.	Sbc(s)I		T=5 L=1	N00

Blue knots in a very small inner ring give the appearance here of a yellow bar crossed with a blue one. The spiral pattern is consistent with an origin at the yellow bar ends. Note the blue knot near a major arm bifurcation at 34,2. The central region of NGC 5426 is visible at the bottom of the field.

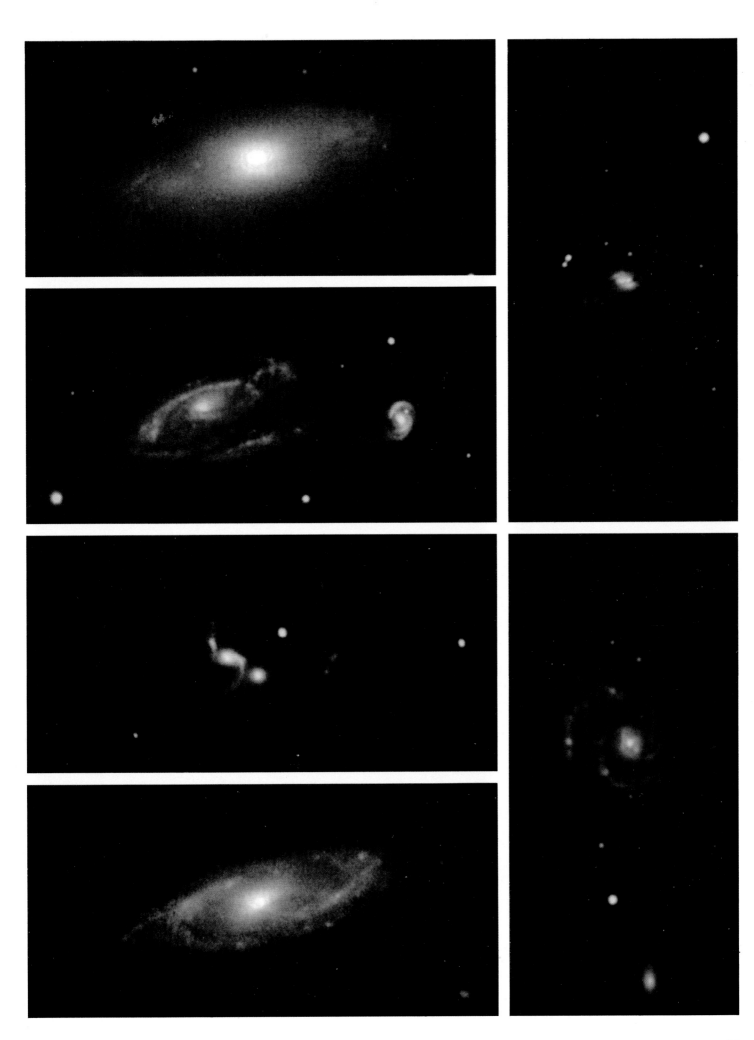

5457
5468
5474
5483
5523
5529

NGC 5457   M 101

1178	6	80 04 21	10.3		McD 0.7m
B=8.2	B−V=0.46	U−B=0.20		V=388
.SXT6..	Sc(s)I		T=6	N08

The small nuclear region is surrounded by an elongated yellow region with the appearance of a short somewhat disorganized bar. Longitudinal dust lanes and a characteristic ring structure commonly associated with bars are also present. Beyond this inner region loose spiral arms wind out but break up into widespread plume structure. Supergiants may be at the threshold of detection.

NGC 5468

1284	6	81 05 07	06.5		LC 1.0m
B=12.2*				V=2764
.SXT6..	Sc(s)II		T=6 L=3	N15

This galaxy is similar to NGC 5457 in color, surface brightness and structure. If its distance is scaled by its apparent size then it is about four times more distant than NGC 5457. Corwin's photometry of the bright star is V=8.36, B−V=0.82 and U−B=0.57.

NGC 5474

783	6	79 03 25	09.2		McD 0.9m
B=11.3	B−V=0.50			V=416
.SAS6P.	Scd(s)IVp		T=6	N18

Note the similarity with NGC 4145 in color and surface brightness particularly in the nuclear region. This galaxy does not appear to be as well organized as NGC 4145, however.

NGC 5523

936b	6	79	05.9		McD 0.9m
B=12.5*				V=1095
.SAS6*.	Sc(s)II−III		T=6 L=4	N18

An apparently reddened system with evidence of dust but no visible star formation.

NGC 5483

1203	6	81 05 01	07.1		CTIO 0.9m
B=11.6	B−V=0.57			V=1628
.SAS5..	SBc(s)II		T=5 L=3	N00

The yellow bar is crossed with dust lanes and appears to contain at least one blue knot. The spiral arms are sharply defined. A faint arm segment splits from the lower arm. Note the bright blue knot at 35,2.

NGC 5529

784	6	79 03 25	09.5		McD 0.9m
B=12.8	B−V=0.85	U−B=0.22		V=2873
.S..5*/			T=5	N21

A massive dust plane obscures most of the lower half of this nearly edge-on system. Note the box shaped nuclear region. Compare with NGC 4586 et al.
A faint blue galaxy is visible at 07,6.

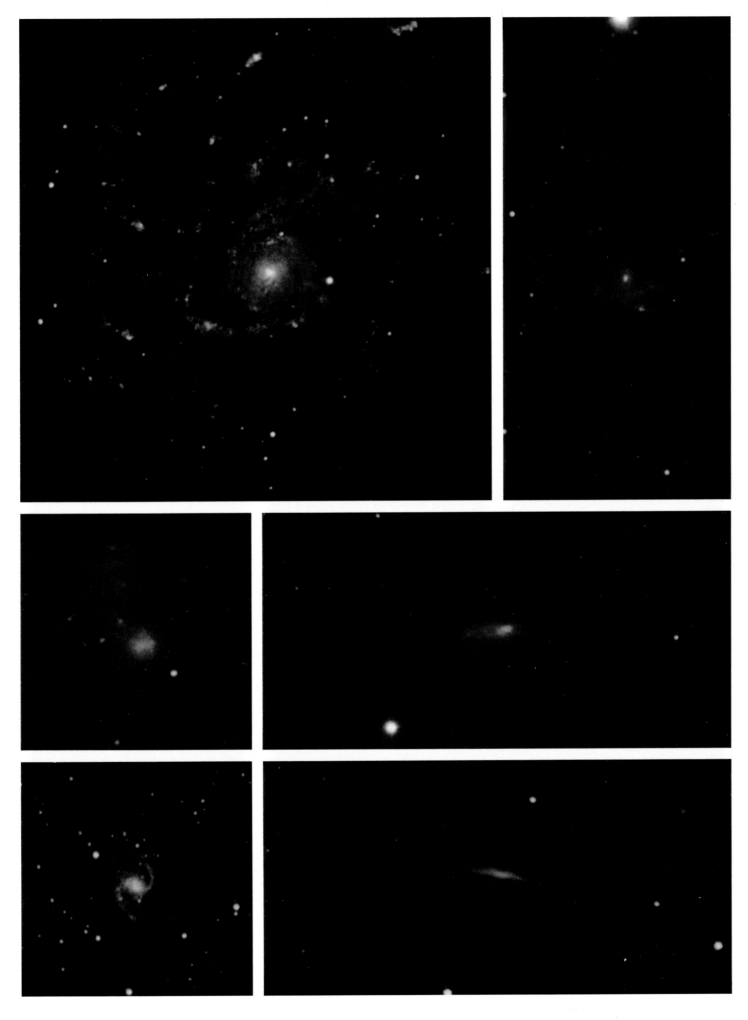

5556
5574
5576
5577
5584
5585
5614
5615
5643

NGC 5556

1285 6 81 05 07 06.8			LC 1.0m
B=12.5	B−V=0.64	U−B=−.10	V=1237
.SXT7..	SBc(sr)II−III		T=7 L=8* N00

The small bar seen here is distinctly green.
Several faint blue knots are visible.

NGC 5574

1083 6 80 03 18 08.2			McD 2.1m
B=13.2	B−V=0.83	U−B=0.28	V=1685
.LBS*?/	S0₁(8)		T=−3 N31

NGC 5574 (right) gives the appearance of a
broad bar structure. Only the old yellow
population is visible in it and in NGC 5576 on
the left.

NGC 5577

1099 6 80 03 19 10.7		McD 2.1m
		V=1460
.SAT4*.		T=4 N33

Two principal spiral arms are defined by loose
groups of faint blue knots. The overall disk
ranges in texture from patchy to smooth. A
background galaxy is visible at 04,3.

NGC 5584

1568 3 82 03 30 10.3		McD 2.7m
B=12.0*		V=1588
.SXT6..	Sc(s)I−II	T=6 N33

The inner spiral pattern is well established
while the outer pattern tends toward plume
structure. This galaxy is rather similar to NGC
3346.

NGC 5585

785 6 79 03 25 09.7		McD 0.9m
B=11.4	B−V=0.50	V=462
.SXS7..	Sd(s)IV	T=7 L=7 N12

A smooth green nuclear region is surrounded
by a few scattered faint blue knots. Is this
perhaps an object with a low metallicity stellar
population? See comments regarding NGC
4449.

NGC 5614

815 3 79 04 24 08.3			McD 2.7m
B=12.5	B−V=0.86	U−B=0.42	V=3966
.SAR2P.	Sa(s) tides		T=2 N00

The system is yellow with some reddening in
the dust lanes. The bright part of the ring on
the right appears to have a slight green
component. Compare with NGC 5929. The
small galaxy at 04,1 is NGC 5615.

NGC 5643

1205 6 81 05 01 07.7			CTIO 0.9m
B=10.7	B−V=0.75	U−B=0.13	V=962
.SXT5	SBc(s)II−III		T=5 L=5 N05

The bar appears relatively well established, but
the ring and spiral structure are somewhat
disorganized. The system is reddened by dust
in our own galaxy.

5668
5669
5676
5678
5701
5739

NGC 5668

1100	6	80 03 19	11.0		McD 2.1m
B=12.1		B−V=0.66			V=1562
.SAS7..		Sc(s)II−III		T=7 L=4 N00	

The nuclear region has a distinct yellow population. Spiral structure is undeveloped in the inner disk, but appears to be slightly better defined in the faint outer regions. Note the sharply defined alignment of faint blue knots near 34,4.

NGC 5669

1570	3	82 03 30	10.7		McD 2.7m
B=12.0*					V=1372
.SXT6..		Sc/SBc(r)I−II		T=6 L=5 N04	

This is another galaxy of the NGC 4145 type: a small inner ring surrounding a short white bar, with spiral arms well developed and defined by blue knots.

NGC 5676

841	3	79 04 25	08.3		McD 2.7m
B=11.6		B−V=0.67			V=2363
.SAT4..		Sc(s)II		T=4 L=3* N31	

This system exhibits a considerably wound spiral structure defined largely by blue knots and intermediate aged features. The right and lower outer region is dominated by plume structure.

NGC 5701

812	6	79 03 26	11.6			McD 0.9m
B=11.8		B−V=0.88		U−B=0.25		V=1493
RSBT0..		(PR)SBa			T=0 L=2 N00	

Only the old yellow population is visible in this photograph.

NGC 5678

887	3	79 04 27	07.7		McD 2.7m
B=12.1*					V=2391
.SXT3..		Sc(s)II−III		T=3 L=3* N08	

This system exhibits a rather high degree of asymmetry. Note the string of blue knots at 07,2 crossing the broad region of plume structure which dominates the right side of the galaxy.

Corwin's photometry of the bright star at 11,8 is V=9.34, B−V=0.84 and U−B=0.58.

NGC 5739

1571	3	82 03 30	10.9		McD 2.7m
B=12.8*					V=5708
.LXR+*.		Sa(s)		T=−1 L=2 N18	

Only a patchy structure probably due to dust is evident here in addition to the reddened old yellow population.

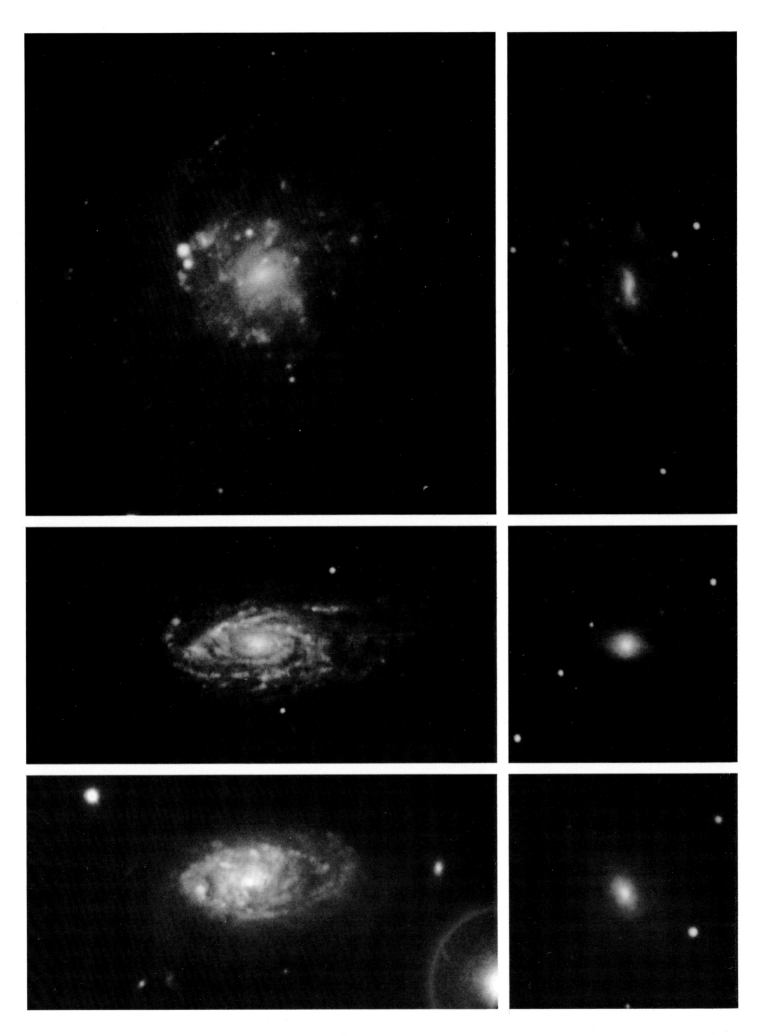

5746
5792
5850
5854
5866
5879
5885

NGC 5746

1183	6	80 06 06 07.6		McD 0.7m
B=11.4	B−V=0.95	U−B=0.38	V=1786
.SXT3$/	Sb(s)		T=3	N25

A massive dust lane produces sufficient extinction in the nuclear region to prevent the central core from being overexposed, and at the same time reddens most of the disk. The 'box' structure is evident in the central bulge. No star formation regions are visible in the disk.

Corwin's photometry of the bright star is V=8.46, B−V=1.40 and U−B=1.65.

NGC 5850

842	3	79 04 25 08.5		McD 2.7m
B=11.7	B−V=0.80	U−B=0.25	V=2354
.SBR3..	SBb(sr)I−II		T=3 L=1 N18

Structure is present in the nuclear region, but its morphology is not clearly discernible. At the ends of the main bar faint enhancements of mixed population including old yellow stars are visible. A very faint outer ring with several faint blue knots is detectable.

NGC 5866 M102

1146	6	80 03 21 11.4		McD 2.1m
B=10.8	B−V=0.85	U−B=0.40	V=874
.LA.+./	S03(8)		T=−1	N22

A dust lane crosses the center slightly inclined to the major axis. Only the old yellow stellar population is evident in this photograph. Corwin's photometry of the bright star is V=11.40, B−V=0.51 and U−B=−.01.

NGC 5792

1260	6	81 05 06 07.8		LC 1.0m
B=11.8	B−V=0.80	U−B=0.18	V=1974
.SBT3..			T=3	N32

Reddening due to dust is evident near the nuclear region. The disk appears patchy and green with several weak blue knots. No spiral pattern is evident.

NGC 5854

1592	3	82 03 31 10.9		McD 2.7m
B=12.6	B−V=0.86	U−B=0.0	V=1632
.LBS+./	Sa		T=−1 L=2 N18

This system which has some of the appearance of a 'box' nucleus also has the appearance of a bar seen nearly end-on. It is almost identical to NGC 4526. See also NGC 7332.

NGC 5879

925b	6	79 05.9		McD 0.9m
B=12.1	B−V=0.60	U−B=0.04	V=1019
.SAT4*$	Sb(s)II		T=4 L=4 N04

A mixed stellar population and dust are evident here, although discrete blue knots are not visible.

NGC 5885

1261	6	81 05 06 08.4		LC 1.0m
B=12.2	B−V=.52	U−B=−.09	V=1964
.SXR5..	SBc(s)II		T=5 L=8* N00

The green nucleus of this very low surface brightness galaxy is essentially all that can be seen here, although in the original print the galaxy is faintly visible over most of the field.

5905
5907
5921
5929
5930
5964

NGC 5905

926b	6	79 05.9		McD 0.9m
B=12.3*				V=3386
.SBR3..		SBb(rs)I		T=3 L=1* N27

A dust lane is visible in the well established yellow bar. A prominent blue knot is situated at the point of contact of this dust lane with the spiral arm. This galaxy is very similar to NGC 5921, below right.

NGC 5929

818	3	79 04 24 09.3		McD 2.7m
				V=2696
.S..2*P			T=2	N00

Both galaxies of this interacting pair are dominated by old yellow stellar populations. NGC 5930, the larger of the two, has the best defined spiral arms of any galaxy comprised of a pure yellow population seen in this atlas. Furthermore, these spiral features are quite bright. Consider also the other galaxies in interacting systems which have smooth bright green spiral features, such as NGC 5394 and NGC 5257. From all of these it would appear that after a massive burst of star formation activity resulting from tidal interaction star formation ceases abruptly and permanently, and the system evolves prematurely into a yellow population. Apparently this process proceeds rapidly enough for some galaxies to exist which have become yellow but which still retain a sharply defined spiral structure as seen here.

NGC 5907

759	6	79 03 24 09.6		McD 0.9m
B=11.0		B−V=0.77		V=780
.SAS5*/		Sc(on edge)		T=5 L=3* N04

This very flat system contains sufficient dust to cause noticeable overall self reddening at this aspect angle. The yellow inner disk is extensive. The outer disk is dominated by younger stars, as evidenced by a number of blue star formation regions which are visible despite the dust. The dark blue spot at 06,2 is a plate defect.

NGC 5921

890	3	79 04 27 08.2		McD 2.7m
B=11.4		B−V=0.63	U−B=0.02	V=1503
.SBR4..		SBbc(s)I−II		T=4 L=2 N18

An excellent example of a barred spiral with a blue ring. Star formation is also visible in the bar. The ring is well formed and already contains some evolved (yellow) stars in the probable vicinity of the libration velocity nodes. The arms are short and faint, with most of the light coming from the star formation regions. Several blue knots are visible in the apparent inter-arm areas, most of which appear to be associated with faint plume structure. The spiral structure is similar to that of NGC 613 which also exhibits star formation in the corresponding inter-arm regions. These galaxies are at rather similar evolutionary stages. See the discussion of NGC 151 and NGC 936.
Corwin has obtained photometry of three stars in this field: for the bright star at the upper right edge of the field V=10.31, B−V=0.40 and U−B=−.01; for the red star at 30,4 V=12.01, B−V=1.24 and U−B=0.97 and for the star at 30,8 V=12.91, B−V=0.58 and U−B=−.02.

NGC 5964

1572	3	82 03 30 11.0	McD 2.7m
.SBT7..			N00

Only the greenish nuclear region and bar of this very low surface brightness galaxy are visible here.

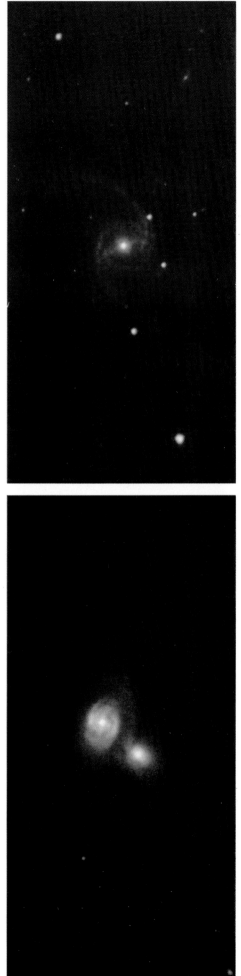

82

5970
5994
5996
6015
6027
6040
6041
6045
6047
6050
6054

NGC 5970

891	3	79 04 27 08.4		McD 2.7m
B=12.1	B−V=0.71	U−B=0.06		V=2061
.SBR5..	SBbc(r)II		T=5	L=4 N01

A broad yellow bar is surrounded by a bright yellow disk containing a ring of blue knots. Faint spiral arms are evident, but most of the visible features outside the ring are plumes.

NGC 5994

889	3	79 04 27 08.0		McD 2.7m
				V=3375
				N02

NGC 5994 and the smaller NGC 5996 (14,4) form an interacting pair similar in some respects to NGC 4490. The bar and bar-end enhancements appear to be well established despite the evident youth of their stellar population. This condition could be a result of of the tidal interaction, or it may be a further illustration of the tendency for stable bar structures to form at an early stage in the evolution of a galactic system.

NGC 6015

791	6	79 03 25 11.4		McD 0.9m
B=11.6	B−V=0.57	U−B=−.06		V=1047
.SAS6..	Sc(s)II−III		T=6	L=3 N07

In this galaxy we see a multi-armed plume dominated spiral with a yellow nuclear region and radially increasing star formation activity.

NGC 6027

906	3	79 04 27 10.5		McD 2.7m
B=13.3	B−V=0.91	U−B=0.32		V=4413
.P.....				N12

Only one galaxy (at 30,2) in the NGC 6027 group has a significant population of young stars. The only other galaxy (05,1) in the group to exhibit any star formation activity at all is dominated by the old yellow population.

NGC 6054

822	3	79 04 24 10.6		McD 2.7m
B=15.8	B−V=0.70	U−B=0.5		V=11128
PSBS3P.			T=3	l=3* N06

NGC 6054 is the galaxy at lower left with the extraordinary blue bar. There are practically no galaxies with bars similar to this, although if star formation activity was greater in the bar in NGC 4151 (which has a blue bar structure of similar galactic scale but much fainter) it would be similar in appearance and perhaps in nature. More curious is the incredible compound coincidence of the occurrence, in the same field, of a galaxy (IC 1182 at 02,2) with a thin jet terminating in blue knots (34,5), an object with virtually the same morphological structure as NGC 3561b which occupies the same field as NGC 3561a a galaxy with a blue bar structure similar to NGC 6054. At the moment we are constrained to wonder. Closer examination of IC 1182 also reveals a blue region at the edge of the bright yellow disk nearly opposite to the direction of the jet. Each of these remarkable objects needs to be studied at much higher resolution in order to more fully comprehend the synergistics of this astonishing system.

NGC 6040

819	3	79 04 24 10.0		McD 2.7m
B=15.0	B−V=0.96	U−B=0.0		V=12729
.SXR5*P			T=5*	L=2 N32

This field and the following three illustrate parts of the Hercules Supercluster of galaxies, a group distinct from the Local Supercluster to which most of the other galaxies in this atlas belong.

NGC 6040a is the rather normal appearing spiral with blue disk at top left. The adjacent elliptical is NGC 6040b. The two galaxies at the bottom are NGC 6041a (left) and NGC 6041b. The galaxy at 16,3 is IC 1170. Galaxies with star formation are relatively rare in groups of closely associated galaxies. This is an effect presumably due primarily to gas-stripping collisions, but one which may also be due at least in part to accelerated consumption of star forming materials. See for example NGC 5929, NGC 5257 and others.

NGC 6045

848	3	79 04 25 10.1		McD 2.7m
B=15.0	B−V=0.96	U−B=0.0		V=10046
.SBT5*P			T=5*	L=1* N15

NCG 6045 is the spiral at the bottom of the field. Its long shape is probably due to tidal effects perhaps caused by its small apparent companion galaxy at the tip of the right spiral arm. Individual blue knots are visible in NGC 6045. If the absolute magnitude of the brightest individual blue knot is about −13.0, and its apparent magnitude is about +21.5 then the distance modulus would be about +34.5. Although the supergiant stars offer a much more definitive basis for deriving distance moduli, the brightness of the brightest blue knot in a galaxy may some day become useful as a distance indicator for galaxies ten times more distant than those in which we can detect the supergiants.

The elliptical at 24,2 is NGC 6047, and a third galaxy is visible at 00,6.

NGC 6050

905	3	79 04 27 10.3		McD 2.7m
B=14.7	B−V=0.68	U−B=0.0		V=9639
.SAS5..			T=5	L=2 N06

NGC 6050a (above center) and b (below center) appear in partial juxtaposition, but lack other evidence of actual collision. Both have widespread star formation activity. NGC 6050b has a bar surrounded by an inner blue ring. A third galaxy is superimposed on the right spiral arm of NGC 6050a. It is slightly green in comparison with the nuclear region of NGC 6050a. A fourth galaxy is visible at 24,2. Its nuclear region is yellow, and its disk somewhat blue. A yellow elliptical galaxy is visible at 19,3.

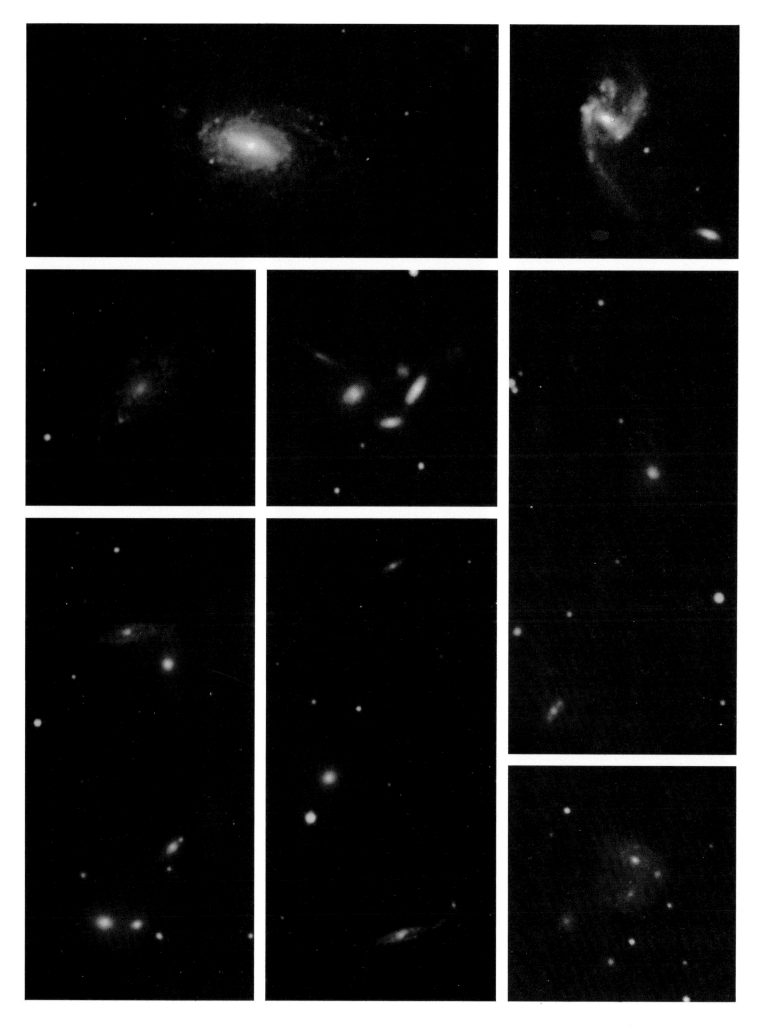

6070
6118
6166
6207
6221
6300
6340
6384
6503

NGC 6070

938b 6 79 04			McD 0.9m
B=12.3	B−V=0.67	U−B=0.10	V=2060
.SAS6..	Sc(s)I		T=6 L=1 N18

The yellow nuclear region is surrounded by a relatively faint but well developed spiral pattern. Several bright blue knots are present in the outer part of the spiral disk.

NGC 6118

1259 6 81 05 06 07.6			LC 1.0m
B=12.4	B−V=0.75	U−B=0.10	V=1618
.SAS6..	Sc(s)I.3		T=6 L=3 N32

The disk appears to be mostly intermediate in age. The spiral pattern is considerably wound and somewhat diffuse. Only one distinct blue knot is visible.

NGC 6166

907 3 79 04 27 10.7			McD 2.7m
B=13.0	B−V=1.06	U−B=0.53	V=9075
.E+2.P.			T=−4 N30

The large galaxy, NGC 6166, is thought to be absorbing the smaller galaxies seen within its envelope. No spiral structure or star formation is visible in any of the galaxies in this field.

NGC 6207

792 6 79 03 25 11.6			McD 0.9m
B=12.1	B−V=0.53	U−B=−.18	V=1066
.SAS5..	Sc(s)III		T=5 L=3* N09

There is an indication of a weak yellow population in the central region of this galaxy. The system is dominated by widespread star formation activity. Several blue stellar objects are visible in the field.

NGC 6221

1206 6 81 05 01 07.8			CTIO 0.9m
B=10.8	B−V=0.78	U−B=0.2	V=1254
.SBS5..	Sbc(s)II−III		T=5 L=4 N24

Nearly lost behind the veil of Milky Way stars this galaxy appears surprisingly little reddened considering its galactic latitude of ten degrees south. Several bright regions of star formation are visible.

NGC 6300

1207 6 81 05 01 08.2			CTIO 0.9m
B=11.13	B−V=0.8	U−B=0.2	V=988
.SBR3..	SBb(s)IIp		T=3 N00

The old yellow bar population extends into the inner part of the ring structure. See discussion accompanying NGC 151 regarding this property. The ring structure is broad and green. Outer spiral structure is not evident. This system is beyond NGC 5921 in evolutionary stage, but the disk has not developed to the stage of NGC 3351. See discussion accompanying NGC 936.

NGC 6340

1574 3 82 03 30 11.4		McD 2.7m
B=11.9	B−V=0.87	V=2146
.SAS0..	Sa(r)I	T=0 L=4 N00

Only the central region of the galaxy is visible in this photograph.

NGC 6384

1538 6 82 03 28 10.4			McD 0.7m
B=11.3	B−V=0.73	U−B=0.5	V=1801
.SXR4..	Sb(r)I		T=4 L=8 N00

Only an old yellow population is clearly visible in this illustration. Very faint spiral features including several faint blue knots are at the threshold of detection.

NGC 6503

1179 6 80 04 21 10.6			McD 0.7m
B=10.9	B−V=0.67	U−B=0.05	V=315
.SAS6..	Sc(s)II.8		T=6 L=5* N01

The inner region is distinctly yellow although it appears to lack a bright nuclear lens. Instead a relatively small nuclear region is visible, possibly somewhat obscured by dust. The outer disk becomes green and contains many regions of star formation. Numerous plumes are present.
Corwin's photometry of the star is V=8.58, B−V=1.25 and U−B=1.43.

NGC 6643

1575	3	82 03 30 11.5			McD 2.7m
B=11.7	B−V=0.68	U−B=−.05			V=1736
.SAT5..	Sc(s)II			T=5	L=2 N00

Plume structure dominates even the inner disk
of this galaxy. Although the structure is
apparently somewhat disorganized, the
brightest regions of star formation occur at
similar distances from the center.

NGC 6744

1208	6	81 05 01 08.7			CTIO 0.9m
B=9.0	B−V=0.65	U−B=0.00			V=519
.SXR4..	Sbc(r)II			T=4	L=3 N09

The yellow bar and the broad patchy ring
structure of intermedate to old age combine
with an outer disk of pure plume structure to
set the character of this galaxy. Bar dust lanes
are broad and very weak. Blue knots are
numerous but not prominent. Supergiant stars
are resolved giving by visual inspection an
approximate distance modulus of about +28.
See NGC 604 and NGC 625.

NGC 6814

939b	6	79 05.9			McD 0.9m
B=12.0	B−V=0.85	U−B=0.22			V=1578
.SXT4..	Sbc(rs)I−II			T=4	L=1 N00

The yellow nuclear region dominates this
spiral. Several blue knots are visible in the faint
spiral structure.

NGC 6872

1361	6	81 10 31 00.6			LC 1.0m
B=12.4	B−V=0.86	U−B=0.49			V=4554
.SBS3P.				T=3	. N18

Long straight spiral arms extend from near the
bar ends. Only one faint blue knot is visible in
these arms. The small galaxy below (IC 4970)
appears to be a spiral, and is probably a
background object.

NGC 6822

1288	6	81 05 07 07.9			LC 1.0m
B=9.3	B−V=0.70	U−B=0.05			V=65
.IBS9..	ImIV−V			T=0	L=8 N33

Only a portion of NGC 6822 is included in
this field. Individual red stars of the giant
branch population are just below the plate
limit.

Two resolved blue knots are visible (04,4 and
28,4). These are outstanding examples of star
formation by interstellar cloud digestion, a self
regenerative process which can be initiated by
the density wave shock process, the several
stochastic model shock processes, or even
spontaneously given the gravitational collapse
of a portion of a compact dust cloud. The
lifetime of the bright blue stars (seen here
individually resolved) is about the same as the
time required for the active star forming region
(located adjacent to the brightest part of the
knot in the dark edge opposite to the tail of
blue stars) to progress a knot diameter (less the
expansion distance) from the star's birth site.
Thus the blue knots remain compact as their
star-forming leading edge eats its way along
the nourishing cloud of gas and dust. See also
NGC 604.

Supergiant stars are detected here a magnitude
or more above the plate limit. The
approximate blue distance modulus by casual
visual estimate of the brightness of the
supergiants is on the order of +26. (see NGC
604 and NGC 625).

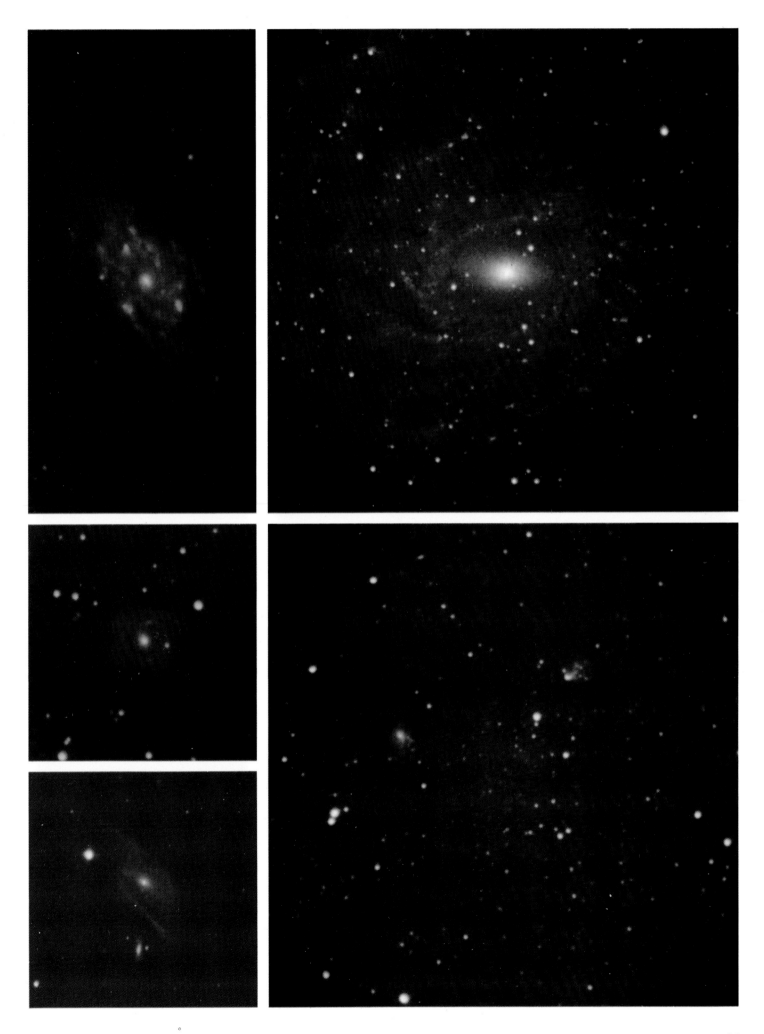

6925
6935
6937
6943
6946
6951

NGC 6925

1307	6	81 10 29 00.8		LC 1.0m
B=12.1	B−V=0.80	U−B=0.18		V=2544
.SAS4..	Sbc(r)I		T=4 L=3	N05

The bright yellow nuclear region appears to have a significant z (out-of-plane) component. More likely it is a bar seen nearly end-on. See NGC 7184. The disk exhibits a spiral pattern of rather uniform color and brightness. Several blue knots are present.

NGC 6935

1333	6	81 10 30 00.4	LC 1.0m
B=12.7	B−V=0.8	U−B=0.2	V=4625
.SXR2..		T=2	N12

NGC 6935 (below) and NGC 6937 both have similar brightness nuclear regions and rather similar rings, but their color properties are noticeably different. The nuclear region of NGC 6935 is somewhat reddish and is surrounded by an apparently reddened disk. The nuclear region of NGC 6937 is distinctly more yellow, almost white, possibly due to a blue contribution from embedded star formation activity. A bright blue knot is seen near the bar end in NGC 6937.

NGC 6943

1308	6	81 10 29 01.4	LC 1.0m
B=11.9	B−V=0.72	U−B=0.1	V=2947
.SXR6*.	Sbc(rs)I	T=6	N18

The nuclear region and inner disk are noticeably red. The spiral features are narrow and relatively disorganized. Some are narrow plumes. Star formation is present in these spiral features.

NGC 6946

940b	6	79 05.9		McD 0.9m
B=9.6	B−V=0.80	U−B=0.05		V=338
.SXT6..	Sc(s)II		T=6 L=1	N09

This galaxy is only twelve degrees from the galactic equator and is probably significantly reddened. The blue knots appear very blue, however. Supergiant stars are probably detected, but are not resolved in the brightest blue knot at 20,5. Any supergiants which are resolved here are easily confused with foreground Milky Way stars.

NGC 6951

324	6	77 11 05 02.0		McD 2.1m
B=12.2	B−V=1.10	U−B=0.41		V=1627
.SXT4..	Sb/SBb(rs)I.3		T=4 L=2*	N00

The bright nuclear region appears to be elongated along an axis at nearly right angles to the bar axis, and may contain star formation. See NGC 4314 and NGC 1365. Prominent dust lanes are visible. The spiral pattern is faint and mostly undetected here.

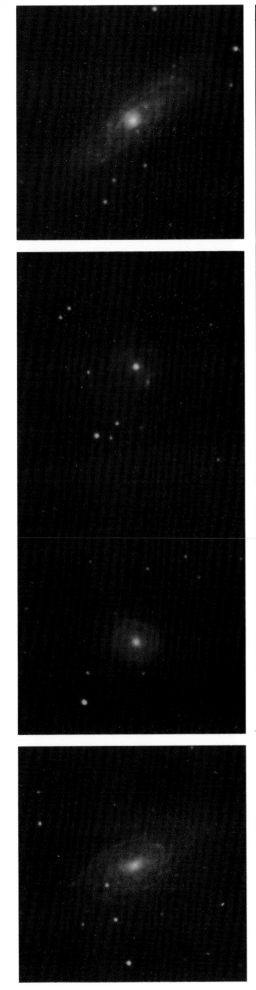

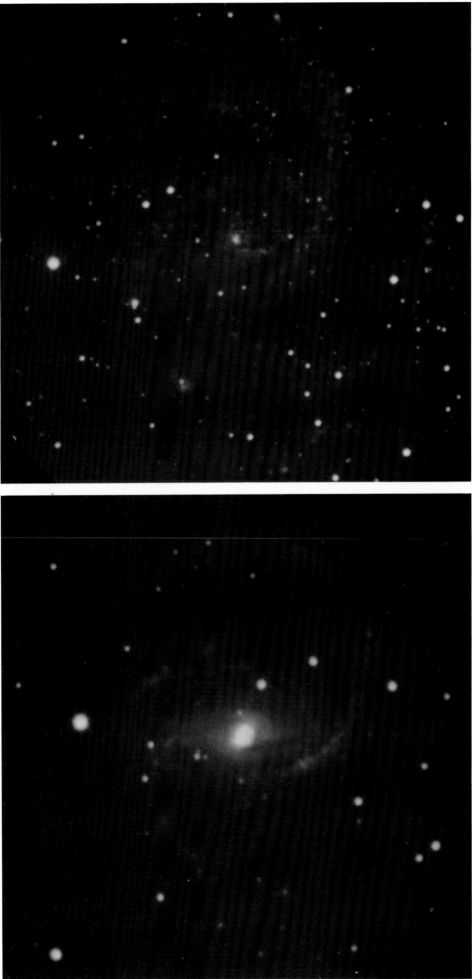

NGC 7038

1334	6	81 10 30 00.7		LC 1.0m
B=12.4*				V=4774
.SAS5*.	Sbc(s)I.8		T=5	N18

A bright yellow nuclear region appears to be accompanied by a faint bar. The inner disk is patchy. Several long spiral features are evident. Two unresolved blue knots are present.

NGC 7049

1363	6	81 10 31 01.2		LC 1.0m
B=11.8	B−V=1.06	U−B=0.60		V=2122
.LAS0..	S0₃(4)/Sa		T=−2	N00

Note the reddening in the dust lane. The disk outside the dust lane appears to be primarily of intermediate population.

NGC 7083

1310	6	81 10 29 02.0		LC 1.0m
B=11.7	B−V=0.60	U−B=0.00		V=2936
.SAS4..	Sb(s)I−II		T=4	L=3 N33

The yellow nuclear region is surrounded by a patchy disk with an evident spiral pattern. Several blue knots are present.

NGC 7090

1262	6	81 05 06 09.5		LC 1.0m
B=11.1	B−V=0.50	U−B=−.05		V=708
.SB.5$/	SBc:(on edge)		T=5	N19

This galaxy may have a white bar. The disk is patchy but somewhat diffuse which is consistent with its color. A faint blue knot is visible just above the bar. There is noticeable reddening in some areas due to dust clouds (note particularly the region near the right end of the bar).

NGC 7096

1311	6	81 10 29 02.4		LC 1.0m
B=12.4	B−V=0.90	U−B=0.40		V=2814
.SAS1..	Sa(r)I		T=1	N00

The old yellow nuclear region dominates this galaxy. Faint, thin, well developed blue spiral arms are present.

NGC 7184

390	6	77 11 06 01.7		McD 2.1m
B=11.7	B−V=0.80	U−B=0.20		V=2722
.SBR5..	Sb(r)II		T=5	L=4 N09

The yellow bar of NGC 7184 is seen nearly end-on. The bar dust lanes are both visible, although the one on the front side is more prominent. The inner ring is well defined and rather green. The outer disk has a disorganized pattern of spiral features. Star formation increases near the outer margin of the disk.

NGC 7124

1312	6	81 10 29 02.8		LC 1.0m
B=13.1				V=4949
.SBT4..	SbcI		T=4	N09

This barred galaxy has a faint blue ring and long narrow blue spiral arms.

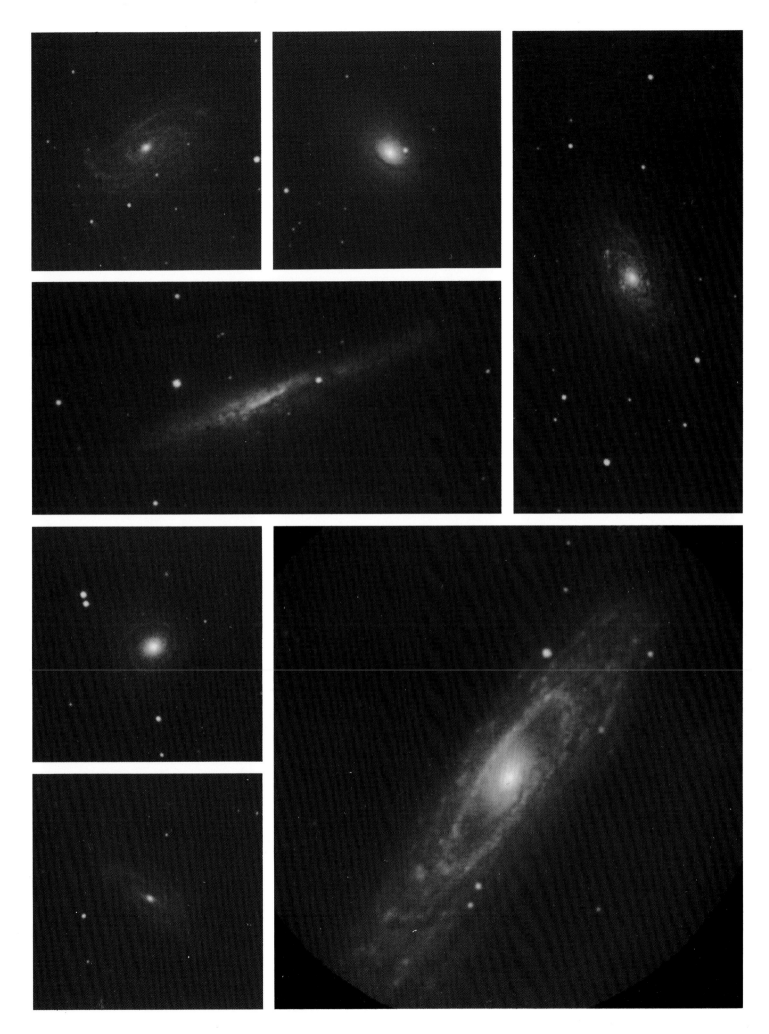

7205
7217
7236
7237
7242
7314

NGC 7205

1336	6	81 10 30 01.3			LC 1.0m
B=11.4	B−V=0.62	U−B=−.03			V=1383
.SXS7*.	Sb(r)II.8		T=7		N20

The nuclear region is yellow and rather small. The spiral pattern is characterized by a single broad rather patchy arm defined almost entirely by diagonal plume features. There is considerable star formation activity.

NGC 7217

327	6	77 11 05 02.4			McD 2.1m
B=11.1	B−V=0.90	U−B=0.39			V=1227
RSAR2..	Sb(r)II−III		T=2	L=3*	N00

Much of the disk is dominated by an old yellow population. Even in the yellow inner disk the surface brightness structure is finely detailed. In the outer disk a broad blue ring is separated from the yellow disk by a similarly broad darker zone. This zone marks a transition between the tightly wound spiral appearance in the inner disk and pure plume structure in the blue ring. A ring is seen in the yellow disk (at radius value 2). This ring appears to be comprised of blue knots, their color having been neutralized by the yellow light. In this case a first wave of star formation has worked its way to the outer region of the galaxy and a second wave has begun to work its way out from the center. See related comments accompanying NGC 4826.

NGC 7236

916	3	79 11 25 01.4			McD 2.7m
B=14.5	B−V=1.04	U−B=0.49			V=8098
.LA.−..			T=−3		N00

NGC 7236 (on the right) and NGC 7237 appear to consist of the purely old yellow type of stellar population

NGC 7314

393	6	77 11 06 02.2			McD 2.1m
B=11.6	B−V=0.62	U−B=−.05			V=1694
.SXT4..	Sc(s)III		T=4	L=2	N27

A small yellow nucleus and bright green disk dominated by plume structure characterize this galaxy. Blue knots are present throughout the disk but tend to occur mostly near its periphery. A long stretch of spiral arm at the top appears to lack star formation activity.

NGC 7242

978	3	79 11 26 02.3			McD 2.7m
B=13.2	B−V=1.12	U−B=0.71			V=5972
.LAR0*.			T=−2		N00

There is no evidence for star formation associated with this highly reddened galaxy. This galaxy has the largest (reddest) B−V value of all the galaxies in the atlas for which measurements are available.

NGC 7317   Stephan's Quintet

473	6	78 01 09 01.9		McD 0.7m
B=14.6	B−V=0.97	U−B=0.45		V=7015
.E.2...			T=−5	N00

NGC 7317 is the small elliptical at lower right of center in this group. The three galaxies at the left exhibit spiral features in this photograph. Blue knots are visible only in NGC 7320 (lower left) which contains at least five distinct regions of star formation. For NGC 7320, V=1042. Note the difference in color between the nuclear region of NGC 7320 and those of the other five galaxies. The center pair is NGC 7318a (left) and NGC 7318b (right). The galaxy at upper left is NGC 7319.

NGC 7339

918	3	79 11 25 01.9		McD 2.7m
B=13.0	B−V=0.86	U−B=0.12		V=1536
.SXS4*$			T=4	N18

The central region is dust reddened. Star formation regions appear to be present in the disk.

NGC 7332

330	6	77 11 05 02.7		McD 2.1m
B=11.8	B−V=0.90	U−B=0.37		V=1451
.L...P/	S02/3(8)		T=−2	N00

This galaxy exhibits a tilted box structure. There is no dust nor evidence for other than the old yellow stellar population.

NGC 7410

1367	6	81 10 31 02.3		LC 1.0m
B=11.3	B−V=0.92	U−B=0.49		V=1634
.SBS1..	SBa		T=1	N32

This barred galaxy appears to have a box nucleus seen from a higher angle of incidence than in any other galaxy in the atlas in which this phenomenon is evident. The stellar population appears to be old, with possibly some stars of intermediate age.

NGC 7412

1314	6	81 10 29 03.3		LC 1.0m
B=11.9	B−V=0.52	U−B=−.02		V=1686
.SBS3..	Sc(rs)I−II		T=3 L=2	N00

A small bright nuclear region is visible together with faint well defined spiral arms containing several bright blue knots. Compare color and color indices with NGC 7418. The color index differences lie along the green-magenta axis of the color-color diagram.

NGC 7418

1338	6	81 10 30 01.8		LC 1.0m
B=12.0	B−V=0.61	U−B=−.10		V=1518
.SXT6..	Sc(rs)I.8		T=6 L=3	N00

The nuclear region is bright and compact, while the bar is unusually red. The spiral pattern is definite, but broken and patchy. The ring contains several very bright blue knots.

NGC 7424

1315	6	81 10 29 03.6		LC 1.0m
B=10.9	B−V=0.50	U−B=−.15		V=850
.SXT6..	Sc(s)II.3		T=6 L=3	N04

Only the small yellow-green bar and several very faint blue knots are visible in this large low surface brightness galaxy.

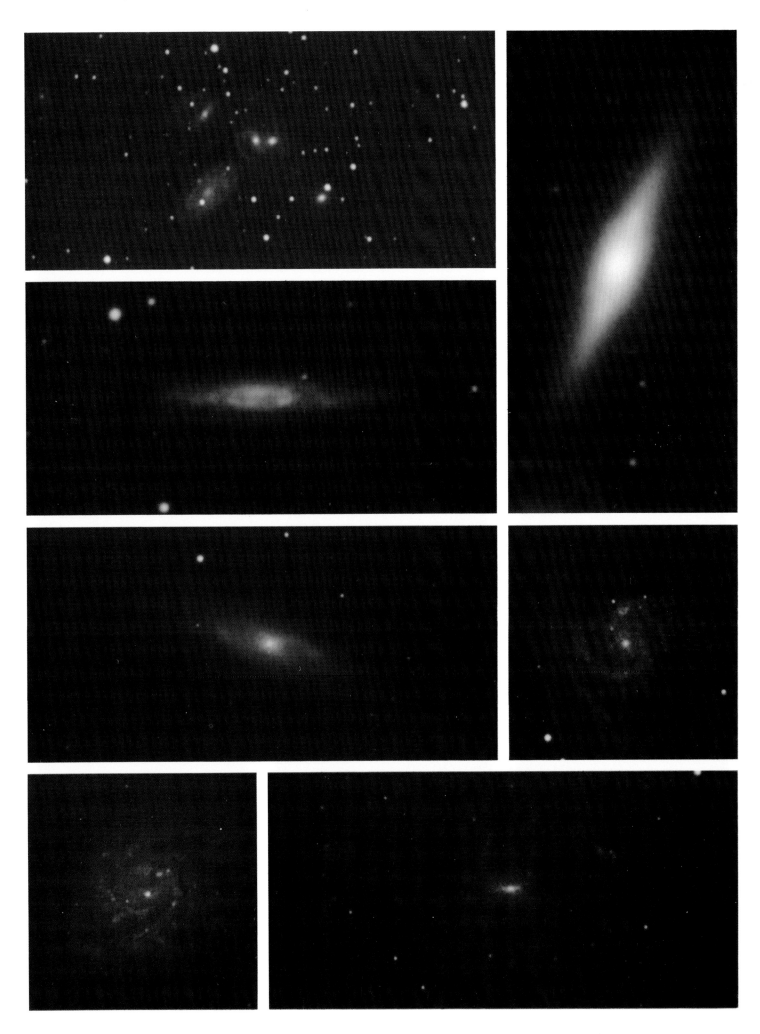

7448
7456
7457
7479
7496
7497
7531

NGC 7448

919   3   79 11 25 02.0   McD 2.7m
B=12.1   B−V=0.50   U−B=−.15   V=2431
.SAT4..   Sc(r)II.2   T=4 L=3* N03

The nuclear region is unusually white. The disk contains many star formation regions. Several very bright regions of star formation are overexposed to a white appearance. Corwin's photometry of the bright star is V=10.32, B−V=0.50 and U−B=0.00.

NGC 7456

1339   6   81 10 30 02.2   LC 1.0m
B=12.2   B−V=0.50   U−B=−.15   V=1205
.SAS6*.   Sc(s)II−III   T=6   N14

The yellow-green nucleus is faint and the disk green and patchy with several small blue knots.

NGC 7457

922   3   79 11 25 02.5   McD 2.7m
B=11.6   B−V=0.89   U−B=0.39   V=790
.LAT−$.   S0₁(5)   T=−3   N00

Only an old stellar population is seen here.

NGC 7479

396   6   77 11 06 02.6   McD 2.1m
B=11.7   B−V=0.70   U−B=0.15   V=2604
.SBS5..   SBbc(s)I−II   T=5 L=1 N00

The nuclear region and bar are comprised of the old yellow stellar population. Dust lanes run along the leading edge of the yellow part of the bar. Star formation regions line the leading edge of the dust lane. This is an outstanding illustration of the prediction of the Non-Linear Density Wave theory which takes gas and dust into account as well as the motion of the stellar density wave. Notice, however, the plume structure, the dust lanes leaving the bar at a large aspect angle and the breakup of the spiral arm at upper left. Additional parameters are required to model such features as these which are so common among galaxies in general.

NGC 7496

1316   6   81 10 29 03.9   LC 1.0m
B=11.6   B−V=0.52   U−B=−.05   V=1443
.SBS3..   SBc(s)II.8   T=3   N00

The nuclear region is bright and contains star formation in a pattern resembling that seen in NGC 5427. One arm segment is undergoing a very active burst of star formation activity with three contiguous bright blue knots.

NGC 7497

399   6   77 11 06 03.0   McD 2.1m
B=12.9   B−V=0.75   U−B=0.15   V=1948
.SBS6..   T=6   N12

The spiral features are broad, diffuse, and rather disorganized. Dust reddening appears to affect the color of this system noticably. A number of star formation regions are visible, but the intermediate aged population tends to dominate the system.

NGC 7531

1340   6   81 10 30 02.4   LC 1.0m
B=11.9   B−V=0.71   U−B=−.03   V=1550
.SAR4..   Sbc(r)I−II   T=4   N27

Surface brightness falls off abruptly outside the bright blue ring of star formation in this galaxy.

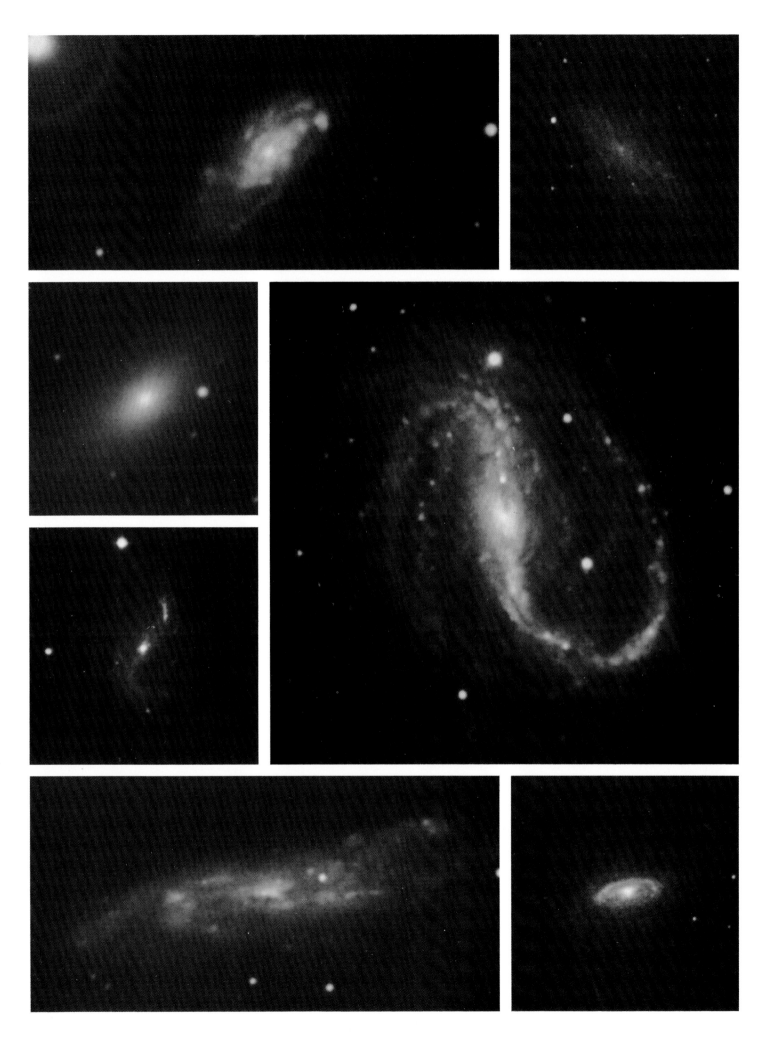

NGC 7541

924   3   79 11 25 02.8   McD 2.7m
B=12.4   B−V=0.73   U−B=0.05   V=2860
.SBT4*P   Sc(s)II   T=4 L=3* N00

This system appears somewhat reddened overall. Blue knots are visible at the upper left and also near the left end of the bright arm segment. The population of this bright arm segment appears to be a mixture of intermediate aged and young stars.

NGC 7552

1410   6   81 11 02 03.7   LC 1.0m
B=11.4   B−V=0.69   U−B=0.10   V=1636
PSBS2..   SBbc(s)I−II   T=2   N00

This somewhat bright system is strikingly reddened in places by dust. A number of blue knots are present. The bar appears to be cut laterally by dust about midway from the center on each side.

NGC 7582

1317   6   81 10 29 04.1   LC 1.0m
B=11.4   B−V=0.77   U−B=0.16   V=1427
PSBS2..   SBab(rs)   T=2   N21

Linear bar dust lanes are evident. A green ring appears foreshortened by the projection angle. Blue knots occur near the interface of dust lane and ring.

NGC 7606

402   6   77 11 06 03.7   McD 2.1m
B=11.5   B−V=0.79   U−B=0.20   V=2361
.SAS3..   Sb(r)I   T=3 L=1 N27

This majestic spiral illustrates the characteristic radial color gradient from yellow to blue with the outwardly decreasing age of the stellar population. Note the two bright blue knots at the edge of the visible luminous disk. Star formation regions are present, however, throughout the entire spiral pattern.

NGC 7678

923   3   79 11 25 02.7   McD 2.7m
B=12.8   B−V=0.65   U−B=−.05   V=3695
.SXT5..   SBbc(s)I−II   T=5 L=1 N00

The bar in this system is somewhat short and faint. The spiral pattern is reasonably well developed. One arm is undergoing a tremendous burst of star formation activity.

NGC 7640

455   6   78 01 08 02.4   McD 0.7m
B=11.3   B−V=0.54   U−B=−.10   V=642
.SBS5..   SBc(s)II: T=5 L=3* N09

A very open barred spiral with yellow nuclear region and green bar. Several star formation regions are evident.

NGC 7713

1429   6   81 11 03 05.0   LC 1.0m
B=11.5   B−V=0.5   U−B=−.15   V=662
.SBR7*.   Sc(s)II−III   T=7   N22

A very small yellow nucleus is visible. The disk is patchy without spiral pattern. Numerous blue knots are present.

NGC 7721

925	3	79 11 25 03.0	McD 2.7m
B=12.3	B−V=0.54	U−B=−.02	V=2177
.SAS5..	Sbc(s)II.2	T=5	L=3 N22

The nuclear region is yellow and somewhat
faint. The spiral arms are broad, open and
diffuse with occasional blue knots. The system
appears to be comprised mostly of stars of
intermediate age.

NGC 7741

239b	6	77 10 15 04.3	McD 0.9m
B=11.9	B−V=0.56	U−B=−.13	V=1018
.SBS6..	SBc(s)II.2	T=6	L=3 N05

The bar is white and contains star formation
regions. The spiral pattern is limited primarily
to a broad ring structure. Although the bar is
well established, and the overall system
structure is reasonably well defined, old
population stars make a minor contribution at
best to the overall stellar population mix.
Corwin's photometry of the bright star is
V=9.81, B−V=1.25 and U−B=1.24. See
below.

NGC 7741

336	6	77 11 05 03.5	McD 2.1m
B=11.9	B−V=0.56	U−B=−.13	V=1018
.SBS6..	SBc(s)II.2	T=6	L=3 N05

The smooth green population underlying the
bar and ring region near the bar ends are more
evident in this photograph at higher
resolution. Many individual blue knots are
visible in the ring. A red background galaxy is
visible at 24,2.
Note the reasonable agreement in color and
surface brightness between this and the
photograph above.

NGC 7742

985	3	79 11 26 03.2	McD 2.7m
B=12.2	B−V=0.71	U−B=−.02	V=1818
.SAR3..	Sa(r!)	T=3	N00

This galaxy has a large smooth yellow outlying
disk. Within the interior of this yellow disk a
brilliant ring of tremendous star formation
activity is seen. This condition can be
understood as an outstanding example of
second wave star formation. See discussion
accompanying NGC 4826.

NGC 7743

339	6	77 11 05 03.9	McD 2.1m
B=12.2	B−V=0.90	U−B=0.42	V=1995
RLBS+..	SBa	T=−1	N08

This galaxy is comprised entirely of an old
stellar population. The system must therefore
be old, yet the spiral pattern is still evident.
This is perhaps the best illustration in the atlas,
and proof for that matter, of the existence of a
pure stellar density wave, as dealt with in the
Linear Density Wave theory which predicts the
behavior of a purely stellar density wave free
from any effects of interstellar gas and dust.

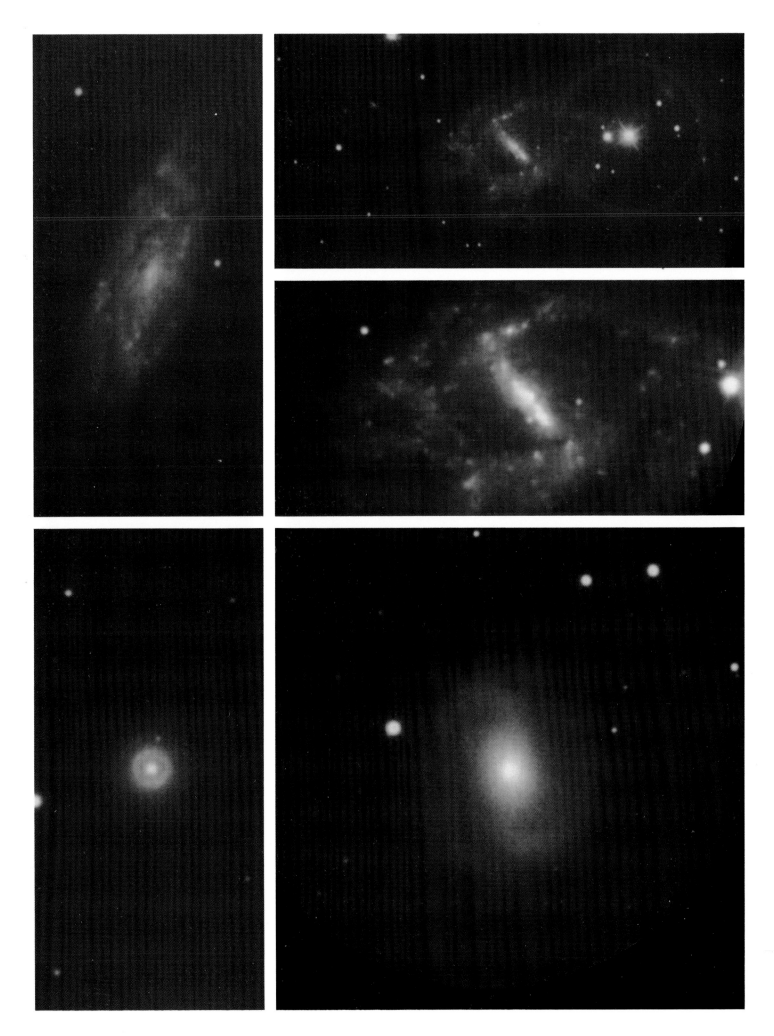

NGC 7753

927 3 79 11 25 03.4				McD 2.7m
B=12.8	B−V=0.80	U−B=0.15		V=5328
.SXT4..			T=4	N05

One arm of NGC 7753 curves down towards the bright galaxy at the bottom of the photograph. The spiral pattern is faint, but contains bright blue knots. The smaller galaxy, NGC 7752, is dominated by star formation activity. Its smooth population appears green rather than yellow.

NGC 7793

1318 6 81 10 29 04.4				LC 1.0m
B=9.7	B−V=0.59	U−B=−.10		V=214
.SAS8..	Sd(s)IV		T=8	L=6 N00

This system, almost entirely lacking in spiral pattern or dynamics, is as good an example of stochastic star formation processes in action as NGC 7479 is of the density wave driven star formation process. Note even here the color gradient from green population in the inner disk to young blue regions in the outer. Dust reddens the disk here in small patches (most visibly near the center).
Supergiant stars are detected just above the plate limit, giving an approximate distance modulus by visual inspection of about +28.

NGC 7757

405 6 77 11 06 04.0				McD 2.1m
B=13.2	B−V=0.45	U−B=−.25		V=3269
.SAT5..			T=5	N33

A very short yellow bar in the nuclear region is surrounded by a relatively green disk with reddened dust lanes. The outer disk comprises areas of plume structure and two very narrow, nearly perfectly aligned, strings of blue knots. The underlying stellar population supporting the density wave is not visible here, only the result; and it is a most remarkable one.
A background galaxy is visible at 04,4.

NGC 7814

408 6 77 11 06 04.4				McD 2.1m
B=11.3	B−V=1.00	U−B=0.50		V=1249
.SAS2*/			T=2	N00

This galaxy with its pure old yellow population and massive dust lanes (which appear to be non co-planar) stands in marked and beautiful contrast to the galaxy NGC 7793 above.

IC 2233

1111	4.5	80 03 21 03.5	McD 2.1m
B=13.5	B−V=0.50	U−B=0.00	V=615
.SB57*.		T=7	N12

An edge-on perspective of a nearly dust free system with considerable star formation activity. There is a slight yellowing near the region of the nuclear bulge.
Corwin's photometry of the bright star is V=10.13, B−V=0.37, U−B=0.14.

A0936−04 Group

1113	4.5	80 03 21 04.0	McD 2.1m
			V=6412
.E.4...		T=−5	N20

The disk of the edge-on galaxy at top left is noticably bluer than the nuclear region, as well as tidally distorted. The only other blue object visible in the entire group is a blue stellar object near the edge of the nuclear region of the barred spiral at the bottom, and which may or may not be associated with that galaxy.

A0705+71

1109	4.5	80 03 21 02.7	McD 2.1m
			V=3309
.5..3*P		T=3*	N00

The central region is reddened mostly by age, and partly by dust. Star formation is occurring along the length of this remarkable 'integral sign' galaxy.

A0708+73

1108	4.5	80 03 21 02.5	McD 2.1m
B=14.2	B−V=0.89	U−B=0.27	V=2900
.RING.A		T=−5P	.N32

This pair is involved in a cataclysmic interaction which is virtually destroying the spiral galaxy.

A0813+70   HoII

507	6	78 03 05 03.9	McD 0.7m
B=11.1	B−V=0.32	U−B=−.33	V=305
.I..9..		T=10 L=8	N00

This illustrates one of the faint dwarf irregular galaxies which may frequent galaxian space for the most part entirely undetected.

A1101+41 Mayall's Object

1124	6	80 03 21 06.2	McD 2.1m
B=15.1	B−V=0.60	U−B=−.22	V=10363
.RING..		T=10R	N00

In this remarkable object the ring appears to have been separated from the rest of the galaxy. Several blue knots are visible in the ring which has no nuclear region at all.

Table of galaxies illustrated in the atlas

No.	Pl.	No.	Pl.	No.	Pl.	No.	Pl.	No.	Pl.	No.	Pl.	No.	Pl.
0013	*1*	1302	*15*	2417	*28*	3185	*40*	4144	*52*	4736	*68*	6040a	*83*
0023	*1*	1309	*15*	2424	*28*	3190	*40*	4145	*52*	4753	*68*	6040b	*83*
0024	*1*	1310	*15*	2427	*28*	3198	*40*	4151	*52*	4754	*68*	6041a	*83*
0045	*1*	1313	*15*	2441	*28*	3223	*40*	4157	*53*	4818	*69*	6041b	*83*
0055	*1*	1316	*16*	2442	*28*	3226	*41*	4162	*53*	4826	*69*	6045	*83*
0068	*1*	1317	*16*	2444	*29*	3227	*41*	4178	*53*	4900	*69*	6047	*83*
0070	*1*	1325	*16*	2445	*29*	3254	*41*	4183	*53*	4902	*69*	6050a	*83*
0071	*1*	1326	*16*	2460	*29*	3256	*41*	4189	*53*	4921	*69*	6050b	*83*
0072	*1*	1326a	*16*	2500	*29*	3261	*41*	4192	*53*	4945	*69*	6054	*83*
0095	*1*	1337	*16*	2507	*29*	3287	*41*	4214	*54*	4965	*70*	6070	*84*
0100	*2*	1353	*16*	2523	*29*	3299	*41*	4216	*54*	4976	*70*	6118	*84*
0127	*2*	1357	*16*	2532	*29*	3310	*41*	4217	*54*	5005	*70*	6166	*84*
0128	*2*	1358	*16*	2535	*29*	3319	*41*	4219	*54*	5033	*70*	6207	*84*
0130	*2*	1365	*17*	2536	*29*	3338	*41*	4236	*55*	5054	*70*	6221	*84*
0134	*2*	1371	*18*	2537	*30*	3344	*42*	4242	*55*	5055	*70*	6300	*84*
0147	*2*	1376	*18*	2541	*30*	3346	*42*	4244	*55*	5068	*71*	6340	*84*
0150	*2*	1380	*18*	2543	*30*	3347	*42*	4254	*55*	5085	*71*	6384	*84*
0151	*2*	1385	*18*	2545	*30*	3351	*42*	4258	*55*	5101	*71*	6503	*84*
0157	*3*	1398	*18*	2549	*30*	3354	*42*	4274	*55*	5102	*71*	6643	*85*
0160	*3*	1415	*18*	2551	*30*	3359	*42*	4293	*55*	5112	*71*	6744	*85*
0185	*3*	1417	*18*	2552	*31*	3368	*43*	4298	*56*	5128	*72*	6814	*85*
0205	*3*	1418	*18*	2563	*31*	3377	*43*	4302	*56*	5161	*73*	6822	*85*
0206*	*3*	1421	*19*	2595	*31*	3379	*43*	4303	*56*	5170	*73*	6872	*85*
0210	*3*	1424	*19*	2598	*31*	3395	*43*	4314	*56*	5172	*73*	6925	*86*
0221	*3*	1425	*19*	2608	*31*	3396	*43*	4321	*57*	5194	*73*	6935	*86*
0253	*4*	1433	*19*	2613	*31*	3423	*43*	4365	*57*	5194	*74*	6937	*86*
0247	*5*	1448	*19*	2614	*31*	3430	*43*	4374	*58*	5195	*73*	6943	*86*
0255	*5*	1453	*19*	2623	*31*	3432	*44*	4382	*58*	5195	*74*	6946	*86*
0278	*5*	1493	*19*	2633	*31*	3437	*44*	4394	*58*	5204	*75*	6951	*86*
0289	*5*	1494	*19*	2639	*32*	3486	*44*	4395	*58*	5230	*75*	7038	*87*
0300	*5*	1507	*20*	2642	*32*	3504	*44*	4402	*59*	5236	*75*	7049	*87*
0309	*6*	1510	*20*	2649	*32*	3511	*44*	4406	*59*	5247	*75*	7083	*87*
0337	*6*	1512	*20*	2654	*32*	3521	*45*	4414	*59*	5248	*75*	7090	*87*
0406	*6*	1515	*20*	2672	*32*	3549	*45*	4429	*59*	5257	*76*	7096	*87*
0428	*6*	1530	*20*	2673	*32*	3556	*45*	4435	*59*	5258	*76*	7124	*87*
0450	*6*	1531	*21*	2681	*32*	3561	*45*	4438	*59*	5266	*76*	7184	*87*
0470	*6*	1532	*21*	2683	*32*	3583	*45*	4449	*60*	5300	*76*	7205	*88*
0474	*6*	1533	*22*	2685	*33*	3596	*45*	4450	*61*	5301	*76*	7217	*88*
0488	*7*	1542	*22*	2712	*33*	3614	*45*	4472	*61*	5363	*76*	7236	*88*
0514	*7*	1549	*22*	2713	*33*	3621	*46*	4485	*61*	5364	*76*	7237	*88*
0520	*7*	1559	*22*	2715	*33*	3623	*46*	4486	*61*	5371	*76*	7242	*88*
0521	*7*	1560	*22*	2742	*33*	3626	*46*	4487	*61*	5377	*77*	7314	*88*
0524	*8*	1566	*22*	2753	*33*	3627	*46*	4490	*61*	5383	*77*	7317	*89*
0578	*8*	1569	*22*	2768	*33*	3628	*46*	4496	*61*	5394	*77*	7318a	*89*
0600	*8*	1617	*22*	2770	*34*	3631	*46*	4501	*61*	5395	*77*	7318b	*89*
0604*	*8*	1618	*22*	2775	*34*	3642	*46*	4504	*62*	5421	*77*	7319	*89*
0613	*8*	1622	*23*	2776	*34*	3664	*47*	4517	*62*	5426	*77*	7320	*89*
0625	*8*	1625	*23*	2782	*34*	3665	*47*	4526	*62*	5427	*77*	7332	*89*
0628	*9*	1635	*23*	2784	*34*	3672	*47*	4527	*62*	5448	*77*	7339	*89*
0660	*9*	1637	*23*	2787	*34*	3675	*47*	4535	*62*	5457	*78*	7410	*89*
0661	*9*	1642	*23*	2798	*34*	3686	*47*	4548	*62*	5468	*78*	7412	*89*
0672	*9*	1659	*23*	2799	*34*	3705	*47*	4550	*63*	5474	*78*	7418	*89*
0678	*9*	1667	*23*	2835	*35*	3717	*47*	4551	*63*	5483	*78*	7424	*89*
0685	*9*	1672	*23*	2841	*35*	3718	*48*	4559	*63*	5523	*78*	7448	*90*
0691	*10*	1688	*23*	2848	*35*	3726	*48*	4564	*63*	5529	*78*	7456	*90*
0694	*10*	1700	*24*	2857	*35*	3733	*48*	4565	*63*	5556	*79*	7457	*90*
0697	*10*	1703	*24*	2859	*35*	3745	*48*	4569	*63*	5574	*79*	7479	*90*
0772	*10*	1720	*24*	2872	*35*	3746	*48*	4579	*63*	5576	*79*	7496	*90*
0784	*10*	1741	*24*	2874	*35*	3748	*48*	4586	*63*	5577	*79*	7497	*90*
0891	*10*	1744	*24*	2903	*36*	3750	*48*	4593	*63*	5584	*79*	7531	*90*
0895	*10*	1752	*24*	2907	*36*	3753	*48*	4594	*64*	5585	*79*	7541	*91*
0908	*11*	1779	*24*	2935	*36*	3754	*48*	4596	*64*	5614	*79*	7552	*91*
0936	*11*	1784	*24*	2936	*36*	3786	*49*	4597	*64*	5615	*79*	7582	*91*
0972	*11*	1792	*25*	2937	*36*	3788	*49*	4605	*64*	5643	*79*	7606	*91*
0986	*11*	1808	*25*	2964	*36*	3808	*49*	4608	*65*	5668	*80*	7640	*91*
1003	*11*	1832	*25*	2967	*36*	3810	*49*	4618	*65*	5669	*80*	7678	*91*
1023	*11*	1888	*25*	2976	*37*	3893	*49*	4621	*65*	5676	*80*	7713	*91*
1042	*12*	1889	*25*	2997	*37*	3896	*49*	4627	*65*	5678	*80*	7721	*92*
1055	*12*	1954	*25*	2998	*37*	3898	*49*	4631	*65*	5701	*80*	7741	*92*
1058	*12*	1964	*25*	3003	*37*	3913	*49*	4633	*65*	5739	*80*	7742	*92*
1068	*12*	2090	*25*	3031	*38*	3917	*50*	4634	*65*	5746	*81*	7743	*92*
1073	*12*	2139	*25*	3032	*38*	3921	*50*	4636	*65*	5792	*81*	7752	*93*
1079	*13*	2146	*26*	3034	*38*	3930	*50*	4643	*65*	5850	*81*	7753	*93*
1084	*13*	2188	*26*	3041	*38*	3938	*50*	4647	*66*	5854	*81*	7793	*93*
1087	*13*	2196	*26*	3054	*38*	3953	*50*	4649	*66*	5866	*81*	7814	*93*
1097	*13*	2207	*26*	3059	*38*	3981	*50*	4651	*66*	5879	*81*	IC 2233	*94*
1156	*14*	2217	*26*	3067	*39*	3992	*50*	4654	*66*	5885	*81*	A0705+71	*94*
1186	*14*	2223	*26*	3077	*39*	4026	*51*	4656	*66*	5905	*82*	A0708+73	*94*
1187	*14*	2256	*26*	3079	*39*	4030	*51*	4657	*66*	5907	*82*	A0813+70	*94*
1201	*14*	2280	*27*	3109	*39*	4038	*51*	4665	*66*	5921	*82*	A0936−04	*94*
1232	*14*	2310	*27*	3113	*39*	4051	*51*	4666	*67*	5929	*82*	A1101+41	*94*
1241	*14*	2336	*27*	3115	*39*	4085	*51*	4676a	*67*	5930	*82*		
1242	*14*	2339	*27*	3124	*39*	4088	*51*	4676b	*67*	5964	*82*		
1255	*14*	2347	*27*	3147	*39*	4094	*51*	4697	*67*	5970	*83*		
1275	*15*	2366	*27*	3162	*39*	4096	*52*	4699	*67*	5994	*83*		
1288	*15*	2389	*27*	3166	*39*	4100	*52*	4710	*67*	5996	*83*		
1291	*15*	2403	*28*	3175	*40*	4123	*52*	4725	*67*	6015	*83*		
1300	*15*	2415	*28*	3184	*40*	4136	*52*	4731	*68*	6027+	*83*		